PHYSICS AND
BEYOND
ENCOUNTERS AND CONVERSATIONS

PHYSICS AND
BEYOND
ENCOUNTERS AND CONVERSATIONS

WERNER HEISENBERG
Translated from the German by Arnold J. Pomerans

HARPER TORCHBOOKS
Harper & Row, Publishers
New York, Hagerstown, San Francisco, London

This Torchbook paperback edition reprints Volume XLII of the WORLD PERSPECTIVES series which is planned and edited by RUTH NANDA ANSHEN. Dr. Anshen's Epilogue to this reprint appears on page 249.

First HARPER TORCHBOOK edition published 1972

STANDARD BOOK NUMBER: 06-131622-9

78 79 80   12 11 10 9 8 7

Contents

	Preface	vii
1	First Encounter with the Atomic Concept (1919–1920)	1
2	The Decision to Study Physics (1920)	15
3	"Understanding" in Modern Physics (1920–1922)	27
4	Lessons in Politics and History (1922–1924)	43
5	Quantum Mechanics and a Talk with Einstein (1925–1926)	58
6	Fresh Fields (1926–1927)	70
7	Science and Religion (1927)	82
8	Atomic Physics and Pragmatism (1929)	93
9	The Relationship between Biology, Physics and Chemistry (1930–1932)	103
10	Quantum Mechanics and Kantian Philosophy (1930–1934)	117
11	Discussions about Language (1933)	125
12	Revolution and University Life (1933)	141
13	Atomic Power and Elementary Particles (1935–1937)	155

14   Individual Behavior in the Face of Political
     Disaster (1937–1941)   165

15   Toward a New Beginning (1941–1945)   180

16   The Responsibility of the Scientist (1945–1950)   192

17   Positivism, Metaphysics and Religion (1952)   205

18   Scientific and Political Disputes (1956–1957)   218

19   The Unified Field Theory (1957–1958)   230

20   Elementary Particles and Platonic Philosophy
     (1961–1965)   237

     Epilogue by Ruth Nanda Anshen   249

Preface

> Now, in what concerns these orations . . . I
> have found it impossible to remember their
> exact wording. Hence I have made each orator
> speak as, in my opinion, he would have done in
> the circumstances, but keeping as close as I
> could to the train of thought that guided his
> actual speech.
>
> —THUCYDIDES

Science is made by men, a self-evident fact that is far too often
forgotten. If it is recalled here, it is in the hope of reducing the
gap between the two cultures, between art and science. The
present book deals with the developments of atomic physics during the past fifty years, as the author has experienced them.
Science rests on experiments; its results are attained through
talks among those who work in it and who consult one another
about their interpretation of these experiments. Such talks form
the main content of this book. Through them the author hopes
to demonstrate that science is rooted in conversations. Needless
to say, conversations cannot be reconstructed literally after several decades. Nor is the book intended as a collection of memoirs.
Instead, the author has freely condensed and sacrificed certain
details; all he wishes to reconstruct is the broader picture. In
these conversations atomic physics does not invariably play the
most important role—far from it. Human, philosophical or political problems will crop up time and again, and the author
hopes to show that science is quite inseparable from these more
general questions.

Many of the dramatis personae are referred to by first name, partly because they are not known to the general public, and partly because the author's relationship to them is best conveyed in that way. Moreover, this should help to avoid the impression that the author is presenting a verbatim report, true in every detail. For that reason there has been no attempt to draw a more precise picture of these personalities; they can, as it were, be recognized only from their manner of speech. Careful attention, however, has been paid to the precise atmosphere in which the conversations took place. For in it the creative process of science is made manifest; it helps to explain how the cooperation of different people may culminate in scientific results of the utmost importance. The author will be most happy, if, in this way, he can convey even to those remote from atomic physics some idea of the mental processes that have gone into the genesis and development of that science, and this despite the fact that he has been obliged to introduce some highly abstract and complex mathematical relations.

And finally, by recalling these conversations, the author has tried to pursue an even wider objective. Modern atomic physics has thrown fresh light on basic philosophical, ethical and political problems. Perhaps it is not too much to hope that this book may help to draw the largest possible circle of people into this vital discussion.

1

First Encounter with the
Atomic Concept (1919–1920)

It must have been in the spring of 1920. The end of the First World War had thrown Germany's youth into a great turmoil. The reins of power had fallen from the hands of a deeply disillusioned older generation, and the younger one drew together in an attempt to blaze new paths, or at least to discover a new star by which they could guide their steps in the prevailing darkness. And so, one bright spring morning, some ten to twenty of us, most of them younger than myself, set out on a ramble which, if I remember correctly, took us through the hills above the western shore of Lake Starnberg. Through gaps in the dense emerald screen of beech we caught occasional glimpses of the lake beneath, and of tall mountains in the far distance. It was here that I had my first conversation about that world of atoms which was to play so important a part in my subsequent life. To explain why a group of young nature lovers, enraptured by the glorious spring landscape, should have engaged in such conversations in the first place, I ought perhaps to point out that the cocoon in which home and school protect the young in more peaceful periods had burst open in the confusion of the times, and that, by way of a substitute, we had discovered a new sense of freedom and did not think twice about offering views on even such subjects as called for much more basic information than any of us possessed.

Just a few steps in front of me walked a fair, tall boy whose parents had once asked me to help him with his homework. A year earlier, at the age of fifteen, this boy had been dragging up ammunition for his father, who was manning a machine gun behind the Wittelsbach Fountain. Those were the days of the Soviet Republic in Munich. Others, including myself, had been working as farm laborers in the Bavarian Highlands. And so a rough life was not entirely alien to us; and we were not afraid to form opinions on the most abstruse topics.

Our talk probably turned to atoms because I was preparing for my matriculation in the summer, and hence liked to discuss scientific subjects with my friend Kurt, who shared my interests and hoped to become an engineer. Kurt came from a Protestant officer's family; he was a good sportsman and an excellent comrade. The year before, when Munich had been surrounded by government troops and our families had long since eaten their last piece of bread, he, my brother and I had gone on a foraging expedition to Garching, right through the front lines, and had returned with a rucksack full of bread, butter and bacon. Such shared experiences make for trust and happy understanding. I now told Kurt that I had come across an illustration in my physics book that made no sense to me at all. It was meant to depict the basic chemical process of two uniform substances combining into a third uniform substance, i.e., into a chemical compound. The processes involved, so the book contended, were best explained by the assumption that the smallest particles, or atoms, of either element, combined into small groups of atoms, called molecules. A carbon dioxide molecule, for instance, was said to consist of an atom of carbon and two atoms of oxygen. It was this process which our book tried to illustrate. And in order to explain why it was that precisely one atom of carbon and two atoms of oxygen formed a carbon dioxide molecule, the artist had furnished the atoms with hooks and eyes, by which they could hang together. I found this approach wholly unacceptable. To my mind, hooks and eyes were quite arbitrary structures whose shape could be altered at will to adapt them to different technical tasks, whereas atoms and their combination into molecules were supposed to be governed by strict natural laws. This, I felt, left no room for such human inventions as hooks and eyes.

"If you do not hold with hooks and eyes—and I think them fairly suspect myself—you should nevertheless try to get at the particular experiences which persuaded the artist to use this type of representation," Kurt contended. "For modern science starts from experience and not from philosophical speculation. Experience is all we have to go by, provided only we have gathered it with due care. As far as I know, chemists have shown that the elementary particles of a chemical compound are always represented in a fixed ratio by weight. That is remarkable in itself. For even if one believes in the existence of atoms, i.e., the characteristic particles of every chemical element, forces of the kind we normally encounter in nature would hardly suffice to explain why a carbon atom invariably and exclusively attracts two oxygen atoms and binds them to itself. Even if we grant the existence of some kind of attractive force between the two types of atom, how can we explain why a carbon atom should never combine with three instead of the usual two oxygen atoms?"

"Perhaps the carbon and oxygen atoms have such shapes that the combination of three is impossible for spatial reasons alone."

"If you assume that, and it seems a plausible enough idea, then you are back with much the same things as the hooks and eyes of your textbook. Perhaps the artist wanted to express just that, for he, too, had no idea what atoms really look like. He simply drew hooks and eyes in order to drive home the point that there are forms which lend themselves to the union of two but never of three oxygen atoms with one of carbon."

"Very well, the hooks and eyes have no real meaning. But you claim that the natural laws responsible for the existence of atoms also endow them with just the form that will ensure the right kind of combination. Unfortunately, neither of us is familiar with that form, nor, for that matter, was the illustrator of the textbook. The only thing we can say is that it is thanks to this form that one carbon atom combines with two rather than three oxygen atoms. The chemists, as the book tells us, have invented the concept of 'valence' for this very purpose. But it remains to be seen whether 'valence' is just a word or a truly useful concept."

"It is probably more than just a word; for in the case of the carbon atom the four valence bonds it is said to have—pairs of which are assumed to join up with the two bonds of each oxygen atom—must somehow be related to its tetrahedral form. There is little doubt, therefore, that the valence concept is based on empirical fact, much more so than we can grasp at the moment."

At this point, Robert joined our conversation; he had been

walking silently beside us, but had obviously been listening. Robert had a thin but strong face, framed by dark hair, and at first sight looked rather withdrawn. He rarely joined in the sort of flighty conversations we were wont to have on our walks, but at night, whenever readings were held in the tent, or at mealtimes when we liked to listen to poetry, we would invariably turn to him, for none of us knew more about German poetry, or, indeed, about the philosophers, than he did. Whenever he recited a poem, he would do so without the least kind of pathos, without any strain, and yet the message of the poet would filter through to even the most sober among us. The quiet way in which he formulated his thoughts, his great composure, forced everyone to listen, and what he said struck us as eminently worth listening to. He had obviously been dissatisfied with our conversation about atoms.

"You science worshipers," he said, "speak ever so glibly about experience, and all of you believe that it leads straight to the truth. But if you really think about it, if you really consider what happens during an 'experience,' you will surely have to revise your opinion. Whatever we say is based on thoughts; only our thoughts are directly known to us. But thoughts are not things. We cannot grasp things directly, we must first transform them into ideas, and then shape these into concepts. What reaches us from the outside is a fairly incoherent mixture of odd sense impressions, and these are by no means directly related to the forms or qualities we perceive a posteriori. If, for instance, we look at a square drawn on a piece of paper, neither our retina nor the nerve cells in our brain register anything like a square. To arrive at that, we must first arrange our sense impressions by an unconscious process that helps to transform them into a coherent, 'meaningful' picture. Only through this transformation, through this fitting together of individual impressions into a 'comprehensible' whole, can we claim to have perceived anything. Hence we ought to inquire more closely into the origin of the pictures on which our ideas are based, determine how they can be grasped by concepts, and how they are related to things. Only then can we make authoritative judgments about the meaning of experience. For ideas are obviously prior to experience; indeed, they are the prerequisite of all experience!"

"But surely the very ideas you are so anxious to distinguish from the objects of perception must spring from experience in their turn? Perhaps not as directly as one might naïvely believe, but indirectly, for instance by means of frequent repetitions of similar groups of sense impressions or by linking the evidence of the different senses?"

"That seems far from certain, and not even particularly convincing. I have recently been reading Malebranche, and I was struck by a reference to this very problem. Malebranche examines three possible ways in which ideas can originate. One you have just mentioned yourself: as they impinge on our senses, objects produce concepts directly in our mind. Now this is a possibility Malebranche rejects, on the grounds that sense impressions differ qualitatively both from things and also from ideas. The second possibility is that the human mind has ideas from the outset, or at least has the power to form these ideas by itself. In that case, sense impressions merely remind us of ideas already present or else impel the mind to form them. There is a third possibility— and this is the one that Malebranche plumps for; that the human mind participates in divine reason. It is linked to God, and it is from God that it derives its conceptual power, the images or ideas with which it can arrange the wealth of sense impressions and articulate them conceptually."

"You philosophers are always quick to introduce theology," Kurt objected. "As soon as things get difficult, you produce the great unknown to get you out of your rut. I, for one, refuse to be put off in this way. Since you yourself have posed the question, just tell me how precisely the human mind gets hold of ideas, in this world, not in the next. For the mind and ideas both exist in this world, do they not? If you refuse to admit that ideas originate in experience, then it is up to you to explain how else they come to be part of the human mind. Are you really suggesting that ideas, or at least the ability to form ideas—through which even a child experiences the world—are inborn? If you do, you must believe that our ideas spring from the experiences of earlier generations. Well, as far as I am concerned, it matters little whether present experiences or those of past generations are responsible."

"No," Robert replied, "that was not at all what I meant. On

the one hand, it is extremely doubtful whether learning, which is the result of experience, can be handed down by hereditary processes. On the other hand, Malebranche's view can be expressed without theological overtones, and so be brought into closer line with modern science. I shall try to do so. Malebranche might easily have said that the same tendencies that provide for visible order in the world, for the existence of chemical elements and their properties, for the formation of crystals, for the creation of life and everything else, may also have been at work in the creation of man's mind. It is these tendencies which cause ideas to correspond to things and which ensure the articulation of concepts. They are responsible for all those real structures that only become split into an objective factor—the thing—and a subjective factor—the idea—when we contemplate them from our human standpoint, when we fix them in our thoughts. Malebranche's thesis has this in common with your conviction that all ideas are based on experience: it grants that the ability to form ideas may well have originated in the course of evolution, in the wake of contacts between living organisms and the external world. However, Malebranche went on to stress that these links cannot be explained away by a chain of causally determined, individual processes. In other words, he argues that here—as in the genesis of crystals or living creatures—we come up against higher morphological structures that elude all attempts to contain them in the conceptual couple: cause and effect. The question whether experience is antecedent to ideas or vice versa is probably no more relevant than the old question about the hen and the egg.

"For the rest, I have no wish to interfere in your conversation. I only wanted to warn you against talking too glibly about experience when dealing with atoms, for it might well turn out that your atoms—which, after all, elude direct observation—are not just things but parts of more fundamental structures which cannot be meaningfully divided into idea and object. I agree, of course, that the hooks and eyes in your textbook must not be taken too literally, or for that matter any of those pictures of atoms that abound in popular writings. Such pictures, which claim to facilitate our understanding, only serve to obscure the real problem. But I think that when you speak about 'atomic forms,' as you did just now, you have to be extremely careful. Only if you give the word 'form' a very wide connotation, use it in more than a purely spatial sense, only if you employ it as loosely as, for instance, I myself have just spoken of 'structure,' can I follow you at least part of the way."

I was suddenly reminded of a fascinating book I had read with keen interest a year earlier, though parts of it had completely eluded me. That book was Plato's *Timaeus*. It, too, contains a philosophical discussion of the smallest particles of matter. Robert's words had just given me my first, vague inkling that it was indeed possible to come to grips with these strange mental constructs. Not that these constructs, which I had previously found quite absurd, had suddenly become plausible—it was just that I had glimpsed a path that might lead to them.

To explain why I was so strongly reminded of the *Timaeus* at just that moment, I must briefly recall the circumstances under which I had read the book. In the spring of 1919, Munich was in a state of utter confusion. On the streets people were shooting at one another, and no one could tell precisely who the contestants were. Political power fluctuated between persons and institutions few of us could have named. Pillage and robbery (I was burgled myself) caused the term "Soviet Republic" to become a synonym of lawlessness, and when, at long last, a new Bavarian government was formed outside Munich, and sent its troops into the city, we were all of us hoping for a speedy return to more orderly conditions. The father of the boy whom I had been coaching took command of a company of volunteers, anxious to play their part in the recapture of the city. He asked his son's friends, all of whom were familiar with the locality, to act as guides to the advancing troops. And so we were assigned to Cavalry Rifle Command No. 11, with headquarters in the Theological Training College opposite the university. Here I did my military service, or rather here all of us led a fairly wild and adventurous life. There were no lessons, as so often before, and many of us were anxious to use this freedom to take a fresh look at the world. Most of the boys with whom I later went hiking around Lake Starnberg were somehow or other engaged in the fighting. Our adventures were over after a few weeks; then the shooting died down and military service became increasingly monotonous.

Quite often it happened that, after spending the whole night on guard in the telephone exchange, I was free for a day, and in order to catch up with my neglected school work I would retire to the roof of the Training College with a Greek school edition of Plato's Dialogues. There, lying in the wide gutter, and warmed by the rays of the early morning sun, I could pursue my studies in peace, and from time to time watch the quickening life in the Ludwigstrasse below.

One such morning, when the light of the sun was already flooding across the university buildings and the fountain, I came to the *Timaeus,* or rather to those passages in which Plato discusses the smallest particles of matter. Perhaps that section captured my imagination only because it was so hard to translate into German, or because it dealt with mathematical matters, which had always interested me. In any case, I worked my way laboriously through the text, even though what I read seemed completely nonsensical. The smallest particles of matter were said to be right-angled triangles which, after combining in pairs into isosceles triangles or squares, joined together into the regular bodies of solid geometry: cubes, tetrahedrons, octahedrons and icosahedrons. These four bodies were said to be the building blocks of the four elements, earth, fire, air and water. I could not make out whether these regular bodies were associated with the elements merely as symbols—for instance, the cube with the element earth so as to represent the solidity and balance of that element—or whether the smallest parts of the earth were actually supposed to be cube-shaped. In either case, the whole thing seemed to be wild speculation, pardonable perhaps on the ground that the Greeks lacked the necessary empirical knowledge. Nevertheless, it saddened me to find a philosopher of Plato's critical acumen succumbing to such fancies. I looked for a principle that might help me to find some justification for Plato's speculation, but, try though I might, I could discover none. Even so, I was enthralled by the idea that the smallest particles of matter must reduce to some mathematical form. After all, any attempt to unravel the dense skein of natural phenomena is dependent upon the discovery of mathematical forms, but why Plato should have chosen the regular bodies of solid geometry, of all things, remained a complete mystery to me. They seemed to have no explanatory value at all. If I nevertheless continued reading the Dialogues, it was simply to brush up on my Greek. Yet I remained perturbed. The most important result of it all, perhaps, was the conviction that, in order to interpret the material world we need to know something about its smallest parts. Moreover, I knew from textbooks and popular writing that modern science was also inquiring into atoms. Perhaps, later in my studies, I myself might enter this strange world. But the time was not yet.

Meanwhile, my uneasiness continued, though perhaps it was only part of the general disquiet that had seized all German youth. I kept wondering why a great philosopher like Plato should have thought he could recognize order in natural phenomena when we ourselves could not. What precisely was the meaning of that term? Are order and our understanding of it purely time-bound? We had all of us grown up in a world that had seemed well ordered enough, and our parents had taught us the bourgeois virtues underpinning that order. The Greeks and the Romans had known that, at times, it may become necessary to sacrifice one's life for the sake of maintaining an orderly way of life, and the death of many of our own friends and relatives had shown us that such was the way of the world even now. That was only to be expected. But there were many who now claimed that the war had been a crime, a crime, moreover, committed by the very men who had been responsible for maintaining the old European order, who had tried to defend it come what may. The old structure of Europe had been shattered by our defeat. That, too, was nothing special. Wherever there are wars there must also be losers. But did that mean that all the old structures had to be discarded? Was it not far better to build a new and more solid order on the old? Or were those right who, in the streets of Munich, had sacrificed their lives to prevent a restoration of the old ways and who proclaimed a new order, not just for a single nation, but for all mankind, even though the majority of mankind might have no wish to build such an order? Our heads were full of such questions, and our elders were unable to provide the answers.

After my reading of the *Timaeus* and before our walk on the hills around Lake Starnberg I had another experience which was to affect my later thoughts profoundly, and which I shall report before returning to our discussion of the atomic problem. A few months after they had captured Munich, the government troops pulled out again, and we returned to school. One afternoon, I was buttonholed by an unknown boy on Leopoldstrasse: "Have you heard about the Youth Assembly in Prunn Castle next week?" he asked. "All of us intend to be there, and we want you to come as well. The more the merrier. We want to find out for ourselves what sort of future we should build!" His voice had the kind of edge I had not heard before. And so I decided to go to Prunn Castle, and asked Kurt to join me.

It took the train, which was running somewhat sporadically, several hours to bring us to the Lower Altmühl Valley. In early geological times it must have been the floor of the Danube; here the river Altmühl has twisted and bored its way through the Jura Mountains, and the picturesque valley is crowned with old castles reminiscent of the Rhenish scene. We had to cover the last few miles to the castle on foot, and could see large crowds making for the heights from all sides.

Prunn Castle stands sheer on a rock at the edge of the valley. The courtyard, with its central well, was teeming with people. Most of them were schoolboys, but there was a sprinkling of older boys who had suffered all the horrors of war at the front and had returned to a completely changed world. There were many speeches that day, full of the kind of pathos that would ring quite false today. We argued passionately about whether the fate of our own nation mattered more than that of all mankind; whether the death of those who had fallen for their country had become meaningless through defeat; whether youth had the right to fashion life according to its own values; whether inner truth was more important than all the old forms that had been shaping human life for centuries.

I myself was much too unsure to join in the debates, but I listened and once again thought a great deal about the meaning of "order." From the remarks of the speakers it was clear that different orders, however sincerely upheld, could clash, and that the result was the very opposite of order. This, I felt, was only possible because all these types of order were partial, mere fragments that had split off from the central order; they might not have lost their creative force, but they were no longer directed toward a unifying center. Its absence was brought home to me with increasingly painful intensity the longer I listened. I was suffering almost physically, but I was quite unable to discover a way toward the center through the thicket of conflicting opinions. Thus the hours ticked by, while more speeches were delivered and more disputes were born. The shadows in the courtyard grew longer, and finally the hot day gave way to slate-gray dusk and a moonlit night. The talk was still going on when, quite suddenly, a young violinist appeared on a balcony above the courtyard. There was a hush as, high above us, he struck up the first great D minor chords of Bach's Chaconne. All at once, and with utter certainty, I had found my link with the center. The moonlit Altmühl Valley below would have been reason enough for a romantic transfiguration; but that was not it. The clear phrases of the Chaconne touched me like a cool wind, breaking through the mist and revealing the towering structures beyond. There had always been a path to the central order in the language of music, in philosophy and in religion, today no less than in Plato's day and in Bach's. That I now knew from my own experience.

We spent the rest of the night around campfires and in our tents on a meadow above the castle, giving full rein to our romantic and poetic sentiments. The young musician, himself a student, sat near our group and played minuets by Mozart and Beethoven interspersed with old folk songs; I tried to accompany him on my guitar. Otherwise, he proved a very gay young fellow and was reluctant to discuss his solemn rendering of the Chaconne. When pressed, he came back at us with "Do you know the key of the trumpets of Jericho?" "No." "D minor [d-moll] also, of course." "Why?" "Because they d-moll-ished the walls!" He escaped our wrath only by taking to his heels.

That night had slipped into the twilight of memories by the time I went hiking across the hills round Lake Starnberg, and talked about atoms. Robert's references to Malebranche had convinced me that our experience of atoms can only be indirect: atoms are not things. This was probably what Plato had tried to say in his Timaeus, and, seen in this light, his speculations about regular bodies were beginning to make more sense to me. When modern scientists speak about the form of atoms, they must be using the word "form" in its widest sense, i.e., they must be referring to the atom's structure in time and space, to the symmetrical properties of its forces, to its ability to form compounds with other atoms. In all probability, such structures will forever elude our powers of graphic description, if only because they are not an obvious part of the objective world of things. But perhaps they are nonetheless open to mathematical treatment.

At once I wanted to learn more about the philosophical aspects of the atomic problem, and to that end I mentioned Plato's *Timaeus* to Robert. I asked him if he was in general agreement with Plato's belief that all material things consist of small units, that there must be ultimate particles into which all matter can be divided. I gained the impression that he took a rather skeptical view of the whole question.

He confirmed this when he said: "Your whole manner of posing the problem, of going so far beyond the world of direct experience, is quite foreign to me. I feel much closer to the world of human beings, to lakes and forests, than I do to atoms. I know that we can ask what happens if we keep dividing and subdividing matter, just as we can ask whether the distant stars and their planets are inhabited by living beings. But I myself can't say I have much interest in such questions. Perhaps I don't even want to know the answers. I believe we have far more important tasks than that."

"I don't wish to argue with you about the relative importance of our respective tasks," I told him. "I myself have always been fascinated by science, and I know that many serious people are anxious to learn more about nature and her laws. Who knows but that their work may not prove of the utmost importance for the whole of mankind? But that is not what matters to me at the moment. What worries me is this: it looks very much as if—and Kurt has been saying just that—modern developments in science and technology have brought us very close to the point where we can see individual atoms, or at least their effects, where we can start to experiment with them. Admittedly, we ourselves still know very little about the subject, simply because our studies have not taken us far enough. But, if my prognosis is correct, what would you, as a disciple of Malebranche, say about it?"

"I should expect that atoms would, in any case, behave quite differently from the objects of everyday experience. I could imagine that attempts to divide matter even further might lead us to fluctuations and discontinuities from which it would be quite possible to conclude that matter has a grainy structure. But I also believe that the new structures will elude all our attempts to construct tangible images, that they will prove to be abstract expressions of natural laws rather than things."

"But what if we could see them?"

"We shall never be able to see atoms themselves, only their effects."

"That's a poor excuse of an answer. For the same remark applies to things in general. In the case of a cat, too, all you can see is the reflection of light rays, i.e., the effects of the cat, and not the cat itself. And when you stroke its fur, the situation is much the same!"

"I'm afraid I can't agree with you. I *can* see a cat directly, for when I look at it, I can—indeed, I must—transform my sense impressions into a coherent idea. In the case of the cat we come face to face with two aspects: an objective and a subjective one—the cat as a thing and as a notion. But atoms are quite a different matter. Here notion and thing can no longer be separated, simply because the atom is neither the one nor the other."

Kurt joined in the discussion once again. "The two of you are much too learned for me; you make free with philosophical speculations, when, in fact, you should be consulting experience. Perhaps our studies may one day introduce us to experiments about or with atoms, and then we shall know what atoms really are. We shall probably discover that they are just as real as all other things which lend themselves to experiment. For if it is true that all material things consist of atoms, then it follows that atoms must be just as real as material things."

"No," replied Robert, "I think your conclusion is highly questionable. You might just as well say that, because all living beings are made up of atoms, atoms are fully alive. Clearly, that is nonsense. Only the combination of a great many atoms into larger structures endows these structures with their characteristic qualities or properties."

"And so you think that atoms don't actually exist, that they are not real?"

"You exaggerate again. Perhaps what we are arguing about is

not so much our knowledge of atoms as the meaning of the words 'actually' or 'real.' You have mentioned the *Timaeus* and told us that Plato identifies the smallest particles of matter with mathematical forms, with regular bodies. Even if he was wrong in fact—and Plato had no experience with atoms—he could have been right in principle. Would you call such mathematical forms 'actual' or 'real'? If they express natural laws, that is the central order inherent in material processes, then you must also call them 'actual,' for they act, they produce tangible effects, but you cannot call them 'real,' because they cannot be described as *res*, as things. In short, we do not know what words we should use, and this is bound to happen once we leave the realm of direct experience, the realm in which our language was formed in prehistoric times."

Kurt remained unconvinced. "I should like to leave even this decision to experiment. I cannot believe that the human imagination can tell us anything about the smallest particles of matter before crucial experiments have familiarized us with them. Only careful investigations, conducted without any preconceptions, can help us here. That is precisely why I am so skeptical about philosophical generalizations on so difficult a subject. They merely cement mental prejudices and hinder rather than foster true understanding. I sincerely hope that scientists will come to grips with atoms long before you philosophers have."

By now, the rest of the party had lost patience with us. "For God's sake, can't you ever stop your bickering?" one of them pleaded. "If you want to bone up for your examinations, please do so at home. How about a song?" We began to sing, and the bright sound of young voices, the colors of the blossoming meadows, were suddenly much more real than all our thoughts about atoms, and dispelled the fancies to which we had surrendered.

2

The Decision
to Study Physics (1920)

From school I did not go straight on to the university; there was a sharp break in my life. After my matriculation, I went on a walking tour through Franconia with the same group of friends, and then I fell seriously ill and had to stay in bed for many weeks. During my long recuperation, too, I was locked away with my books. In these critical months I came across a work that I found extremely fascinating, though I was unable to understand it fully. The author was the famous mathematician, Hermann Weyl, and the book was entitled *Space, Time and Matter*. It was meant to provide a mathematical account of Einstein's relativity theory. The difficult mathematical arguments and the abstract thought underlying that theory both excited and disturbed me, and, in addition, confirmed me in my earlier decision to study mathematics at the University of Munich.

During the first days of my studies, however, a strange and, to me, most surprising event took place, which I should like to report in brief. My father, who taught Middle and Modern Greek at the University of Munich, had arranged an interview with Ferdinand von Lindemann, the professor of mathematics, famous for his solution of the ancient problem of squaring the circle. I intended to ask permission to attend his seminars, for which I imagined my spare-time studies of mathematics had fully prepared me; but when I called on the great man, in his gloomy first-floor office furnished in rather formal, old-fashioned style, I felt an almost immediate sense of oppression. Before I could utter a word of greeting to the professor, who rose from his chair very slowly, I noticed a little black dog cowering on the desk, and was forcefully reminded of the poodle in Faust's study. The little beast looked at me with undisguised animosity; I was an unwelcome intruder about to disturb his master's peace of mind. I was so taken aback that I began to stammer, and even as I spoke it dawned on me that my request was excessively immodest. Lindemann, a tired-looking old gentleman with a white beard, obviously felt the same way about it, and his slight irritation may have been the reason why the small dog now set up a horrible barking. His master tried to calm him down, but the little beast only grew more hysterical, so that we could barely hear each other speak. Lindemann asked me what books I had recently been reading, and I mentioned Weyl's *Space, Time and Matter*. As the tiny monster kept up his yapping, Lindemann closed the conversation with "In that case you are completely lost to mathematics." And that was that.

Clearly mathematics was not for me. A somewhat wearing consultation with my father ended with the advice that I ought to try my hand at theoretical physics. Accordingly, he made an appointment with his old friend Arnold Sommerfeld, then head of the Faculty of Theoretical Physics at the University of Munich and generally considered one of the most brilliant teachers there. Sommerfeld received me in a bright study with windows overlooking a courtyard where I could see a crowd of students on benches beneath a large acacia. The small squat man with his martial dark mustache looked rather austere to me. But his very first sentences revealed his benevolence, his genuine concern for young people, and in particular for the boy who had come to ask his guidance and advice. Once again the conversation turned to the mathematical studies I had pursued as a hobby while still at school, and to Weyl's *Space, Time and Matter*. Sommerfeld's reaction was completely different from Lindemann's.

"You are much too demanding," he said. "You can't possibly start with the most difficult part and hope that the rest will automatically fall into your lap. I gather that you are fascinated by relativity theory and atomic problems. But remember that this is not the only field in which modern physics challenges basic philosophical attitudes, in which extremely exciting ideas are being forged. To reach them is much more difficult than you seem to imagine. You must start with a modest but painstaking study of traditional physics. And if you want to study science at all, you must first make up your mind whether you want to concentrate on experimental or theoretical research. From what you have told me, I take it that you are much keener on theory. But didn't you do experiments and dabble with instruments at school?"

I said that I used to like building small engines, motors and induction coils. But, all in all, I had never been really at home in the world of instruments, and the care needed in making relatively unimportant measurements had struck me as being sheer drudgery.

"Still, even if you study theory, you will have to pay particular attention to what may appear trivial little tasks. Even those who deal with the larger issues, issues with profound philosophical implications—for instance, with Einstein's relativity theory or with Planck's quantum theory—have to tackle a great many petty problems. Only by solving these can they hope to get an over-all picture of the new realms they have opened up."

"Even so, I am much more interested in the underlying philosophical ideas than in the rest," I said rather bashfully.

But Sommerfeld would have none of this. "You must remember what Schiller said about Kant and his interpreters: 'When kings go a-building, wagoners have more work.' At first, none of us are anything but wagoners. But you will see that you, too, will get pleasure from performing minor tasks carefully and conscientiously and, let's hope, from achieving decent results."

Sommerfeld then gave me a few more hints about my preliminary studies, and said that he might well come up with a little problem connected with recent developments in atomic theory on which I could try my mettle. And it was decided that I would join his classes for the next few years.

This, my first conversation with a scholar who really knew his way about in modern physics, who had personally made important discoveries in a field impinging on both relativity and quantum theory, had a lasting effect upon me. Though his call for care in small details struck me as eminently reasonable—I had heard it often enough from my own father—I felt dejected at the thought that I was still such a long way from the field that really interested me. No wonder that this interview became the subject of many discussions with my friends. I remember one of these particularly well: it bore on modern physics and the culture of our time.

That autumn, I saw a great deal of the boy who had played Bach's Chaconne so magnificently in Prunn Castle. We would meet in the house of our mutual friend, Walter, himself a fine cellist, and practice for a private recital of Schubert's B Major Trio. Walter's father had died at an early age, and his mother had been left to care for her two sons in a large and very elegantly furnished house in Elisabeth Strasse, just a few minutes' walk from my parents' house in Hohenzollern Strasse. The magnificent Bechstein grand in the living room was an added reason for our frequent visits. After we had finished playing, we would often talk deep into the night, and it was on one such occasion that the conversation came round to my proposed studies. Walter's mother wondered why I had not decided to make music my career.

"From the way you play and speak about music, I get the impression that you are much more at home with art than with science and technology, that you prefer the muses to scientific instruments, formulae and machinery. If I am right, why ever have you chosen natural science? After all, the future of the world will be decided by you young people. If youth chooses beauty, then there will be more beauty; if it chooses utility, then there will be more useful things. The decision of each individual is of importance not only to himself but to the whole of mankind."

"I can't really believe that we are faced with that sort of choice," I said rather defensively. "Quite apart from the fact that I probably wouldn't make a very good musician, the question remains in which field one can contribute most. Now I have the clear impression that in recent years music has lost much of its earlier force. In the seventeenth century music was still deeply steeped in the religious way of life; in the eighteenth century came the conquest of the world of individual emotions; in the nineteenth century romantic music plumbed the innermost depths of the human soul. But in the last few years music seems to have quite deliberately entered a strange, disturbed and rather feeble stage of experimentation, in which theoretical notions take precedence over the desire for progress along established paths. In science, and particularly in physics, things are quite different. Here the pursuit of clear objectives along fixed paths—the same paths that led to the understanding of certain electromagnetic phenomena twenty years ago—has quite automatically thrown up problems that challenge the whole philosophical basis of science, the structure of space and time, and even the validity of causal laws. Here we are on *terra incognita*, and it will probably take several generations of physicists to discover the final answers. And I frankly confess that I am highly tempted to play some part in all this."

My friend Rolf, the violinist, demurred. "As far as I can see, your remarks about modern physics apply equally well to modern music. Here, too, the path seems to be clearly mapped. The old tonal barriers are collapsing and we find ourselves on promising virgin soil, with almost complete freedom to choose what sounds and rhythms we like. Hence the musician has every chance of discovering as many riches as the scientist."

Walter now raised several objections of his own. "I don't really know whether 'freedom of expression' and 'promising virgin soil' are necessarily the same thing. At first sight it admittedly looks as if greater freedom must necessarily mean enrichment, wider possibilities; but this I know to be untrue in art, with which I am more familiar than with science. I would think that progress in art takes place in the following way: First a slow historical process transforms the life of men in spite of themselves, and thereby throws up fresh ideas. A few talented artists then try to give these ideas a visible or audible form by wresting new possibilities of expression from the material with which they work—from colors or musical instruments. This interplay or, if you like, this struggle between the expressive content and the limitations of the expressive medium is, I think, a *sine qua non* of the emergence of real art. If the limitations of the expressive medium were taken away—if in music, for instance, we could produce any sounds we liked—then the struggle would be over, and the artist's effort would reach into a void. For that reason I am skeptical about too much freedom.

"In science," Walter continued, "a continuous flow of new experiments is made possible by new techniques; there are new experiences and as a result new contents may be produced. Here the means of expression are the concepts by which the new ideas are grasped and made explicit. For instance, I have read that Einstein's relativity theory, which interests you so much, was born from the failure of certain experiments designed to demonstrate the motion of the earth through space by means of the interference of light rays. When this demonstration misfired, it became clear that the new results, or, what amounts to the same thing, the new ideas, called for an extension of the means of expression, i.e., of the conceptual system proper to physics. Quite likely, no one anticipated that this would demand radical changes in such fundamental concepts as space and time. It was Einstein's great achievement to appreciate before anyone else that the ideas of space and time were not only susceptible to change but, in fact, had to be changed.

"What you have said about recent developments in physics could reasonably be compared with developments in music in the middle of the eighteenth century. At that time, a gradual historical process had led to a growing awareness of the emotional world of the individual—as all of us know from Rousseau and later from Goethe's *Werther*—and it was then that the great classicists—Haydn, Mozart, Beethoven and Schubert—succeeded in extending the means of expression and so discovered the musical language needed for depicting this emotional world. In modern music, on the other hand, the new contents appear to be highly obscure and implausible, and the plethora of possible expressions fills me with deep forebodings. The path of modern music seems to be determined by a purely negative postulate: the old tonality has to be discarded because we believe that its powers have been exhausted, and not because there are new and more forceful ideas which it is incapable of expressing. Musicians are entirely in the dark about the next step; at best they grope their way forward. In modern science the questions are clearly posed, and the task is to find the right answers. In modern art, however, even the questions are uncertain. But perhaps you had best tell us a bit more about the new fields you intend to explore in the world of physics."

I tried to convey what little bits of knowledge I had gleaned during my illness, mainly from popular books on atomic physics.

"In relativity theory," I told Walter, "the experiments you have mentioned, together with other experiments, caused Einstein to discard the prevailing concept of simultaneity. That in itself was exciting enough. Every one of us thinks that he knows precisely what the word 'simultaneous' means, even if it refers to events that take place at great distances. But we are mistaken. For if we ask how one determines whether two such events are, in fact, simultaneous and then evaluates the various means of verification by their results, nature herself informs us that the answers are not at all clear-cut but depend on the observer's state of motion. Space and time are therefore not independent of each other, as we previously believed. Einstein was able to express the 'new' structure of space and time by means of a simple and coherent mathematical formula. While I was ill, I tried to probe into this mathematical world, which, as I have since learned from Sommerfeld, has already been opened up fairly extensively and has therefore ceased to be unexplored territory.

"The most interesting problems now lie in a different field, in atomic physics. Here we come face to face with the fundamental question why the material world manifests ever-recurring forms and qualities—why, for example, water with all its characteristic properties is invariably reproduced during the melting of ice, the condensation of steam or the combustion of hydrogen. This has been taken for granted in physics, but has never been fully explained. Let us suppose that material bodies—in our case, water —are composed of atoms. Chemistry has long made successful use of this idea. Now, the Newtonian laws we were taught at school cannot tell us why the motions of the particles involved should be as stable as they, in fact, are. Only quite different natural laws can help us to explain why atoms should invariably rearrange themselves and move in such a way as to produce the same substances with the same stable properties. We first caught a glimpse of these laws twenty years ago, in Planck's quantum theory. Since then, the Danish physicist, Niels Bohr, has combined Planck's theory with Lord Rutherford's atomic model. In so doing, he was the first to throw light on the curious stability of atoms which I have just mentioned. But Sommerfeld believes that in this sphere we are still a long way from a clear understanding of the ways of nature. Here we have a vast unexplored field, in which new relationships may be discovered for decades to come. By the appropriate reformulation of natural laws and with correct new concepts we might, for instance, be able to reduce the whole of chemistry to atomic physics. In short, I firmly believe that in atomic physics we are on the track of far more important relations, far more important structures, than in music. But I freely admit that 150 years ago things were the other way round."

"In other words," Walter asked, "you believe that anyone concerned with cultural progress must necessarily make use of the historical possibilities of the age in which he lives? That, if Mozart had been born in our day, he, too, would be writing atonal and experimental music?"

"Yes, I suspect just that. If Einstein had lived in the twelfth century, he would not have been able to make important scientific discoveries."

"Perhaps it is wrong to keep bringing up such great men as Mozart and Einstein," Walter's mother said. "Few individuals get the chance to play such decisive roles. Most of us must content ourselves with working quietly in a small circle, and ought to ask simply whether playing Schubert's B Major Trio is not more satisfactory than building instruments or writing mathematical formulae."

I agreed that I myself had quite a few qualms and mentioned Sommerfeld's quotation from Schiller: "When kings go a-building, wagoners have more work."

"We all feel the same way about it," Rolf declared. "Those of us who want to become musicians have to take infinite pains to master their instruments, and even then can only hope to play pieces that hundreds of better musicians have played much more proficiently. And you yourself will have to spend long hours with instruments that others have built much more competently, or retrace the mathematical thoughts of the masters. True, when all this has been done, the musical wagoners among us are left with no small sense of achievement: constant intercourse with glorious music and the occasional delight of a particularly successful interpretation. Likewise, you scientists will occasionally manage to interpret a relationship just that little bit better than anyone before you, or determine a particular process more accurately than your predecessors. But none of us ought to count on the fact that he will be doing trail-blazing work, that he will make decisive discoveries. Not even when he works in a field where a great deal of territory has still to be opened up."

Walter's mother, who had been listening attentively, now said something, more to herself than to us, as if she were trying to formulate her thoughts as she spoke:

"The parable of the kings and the wagoners may have quite a different import. Of course, superficially it looks as if the glory is entirely the kings', as if the wagoners' work were purely subsidiary and unimportant. But perhaps the very opposite is true. Perhaps the kings' glory rests on the work of the wagoners, on the fact that the wagoners have put in many years of laborious effort, reaping joy and success. Perhaps men like Bach or Mozart are kings of music only because, for two long centuries, they have offered so many lesser musicians the chance of reinterpreting their thoughts with love and conscientious attention to detail. And even the audience participates in this careful work as it hears the message of the great musicians.

"If you look at historical developments—in the arts no less than in the sciences—you will find that every discipline has long periods of quiescence or of slow growth. Even during these periods, however, the important thing is careful work, attention to detail. Everything that is not done with utter devotion falls into oblivion and, in fact, does not deserve to be remembered. And then, quite suddenly, this slow process, in which general historical developments introduce changes in the contents of a particular discipline, opens up new possibilities, quite unexpected contents. Talented men feel an almost magical attraction for the process of growth they can sense at work here, and so it happens that, within a few decades, a relatively small region of the world will produce major works of art or scientific discoveries of the greatest importance. In the late eighteenth century, for instance, classical music poured forth from Vienna; in the fifteenth and sixteenth centuries painting had its heyday in the Netherlands. True, great men are needed to express the new spiritual contents, to create the forms in which further developments can be molded, but they do not actually produce these new contents.

"Of course, it is quite possible that we are on the threshold of an exceptionally fruitful scientific epoch, in which case it would be wrong to dissuade any young man from participating in it. It seems unlikely that important developments will take place in more than one branch of art or science at one time; we ought to be grateful enough if it happens in any one area, if we can share in its glory either as bystanders or as active participants. It would be foolish to expect more. That is precisely why I find popular attacks on modern art—be it painting or music—so unjust. Once music and the plastic arts had solved the great problems posed to them in the eighteenth and nineteenth centuries, there just had to be a more restful period, in which much of the old could be preserved and new things were tested by trial and error. To compare modern compositions with the finest achievements of the great epoch of classical music seems utterly unfair. Perhaps we ought to finish the evening with the slow movement of Schubert's B Major Trio. Let's see how well you can play it."

We did as we were asked, and from the way in which Rolf played the somewhat melancholic C major figures in the second part of the movement, I could sense how sad he was at the thought that the great epoch of European music might be gone forever.

A few days later, when I walked into the hall where Sommerfeld usually gave his lectures, I spotted a dark-haired student with a somewhat secretive face in the third row. Sommerfeld had introduced us during my first visit and had then told me that he considered this boy to be one of his most talented students, one from whom I could learn a great deal. His name was Wolfgang Pauli, and for the rest of his life he was to be a good friend, though often a very severe critic. I sat down beside him and asked him if, after the lecture, I might consult him about my preparatory studies. Sommerfeld now entered the hall, and as soon as he started to address us Wolfgang whispered in my ear: "Doesn't he look the typical old Hussar officer?" After the lecture, we went back to the Institute of Theoretical Physics, where I asked Wolfgang two questions. I wanted to know how much experimental work had to be done by someone interested chiefly in theory, and what he thought of the respective importance of relativity and atomic theory.

"I know," Wolfgang told me in reply to my first question, "that Sommerfeld lays great stress on experimental studies, but I myself am not cut out for them; I hate the whole business of handling instruments. I quite agree that physics is based on experimental results, but once these results have been obtained, physics, at least modern physics, becomes much too difficult a subject for most experimental physicists. This is probably so because the sophisticated instruments of modern physics take us into realms of nature that cannot be adequately described with everyday concepts. We are forced to employ an abstract kind of mathematical language and one that presupposes a considerable amount of training in modern mathematics. It is a sad fact but true that we all have to specialize. I find abstract mathematical language quite easy, and hope to put it to good use in my work. Needless to say, I realize that some knowledge of the experimental side is absolutely essential. The pure mathematician, however good, understands nothing at all about physics."

I then repeated my conversation with old Lindemann, and told Wolfgang about his black lap dog and his reaction to my reading Weyl's *Space, Time and Matter*. My report obviously caused Wolfgang the greatest amusement.

"That's just what I would have expected," he said. "Weyl really does know a lot about relativity theory, and for Lindemann such knowledge is enough to disqualify anyone from bearing the title of serious mathematician."

As to the respective importance of relativity and atomic theory, Wolfgang had this to say: "The so-called special theory of relativity is now a closed chapter; you simply have to learn it and use it like any other theory in physics. Nor is it of particular interest to anyone anxious to make new discoveries. However, the general theory of relativity, or, what comes to much the same thing, Einstein's theory of gravitation, is still wide-open. But it is rather unsatisfying in that, for each experiment, it will give you a hundred pages of theory with the most complicated mathematical derivations. No one can really say whether the whole thing is correct. Nevertheless it opens up new possibilities of thought, and for that reason must be taken seriously. I have recently written a fairly lengthy article on the general theory of relativity; perhaps that is one of the reasons why I find atomic theory so much more interesting.

"In atomic physics we still have a wealth of uninterpreted experimental results: nature's evidence in one place seems to contradict that in another, and so far it has not been possible to draw an even halfway coherent picture of the relationship involved. True, Niels Bohr has succeeded in associating the strange stability of atoms with Planck's quantum hypothesis—which has not yet been properly interpreted either—and more recently Bohr is said to have given a qualitative explanation of the periodic system of the elements and of their chemical properties. But I can't for the life of me see how he could have done so, seeing that he, too, is unable to get rid of the contradictions I have mentioned. In other words, everyone is still groping about in a thick mist, and it will probably be quite a few years before it lifts. Sommerfeld hopes that experiments will help us to find some of the new laws. He believes in numerical links, almost in a kind of number mysticism of the kind the Pythagoreans applied to the harmony of vibrating strings. That's why many of us have called this side of his science 'atomysticism,' though, as far as I can tell, no one has been able to suggest anything better. Perhaps it's much easier to find one's way if one isn't too familiar with the magnificent unity of classical physics. You have a decided advantage there," Wolfgang added with a malicious grin, "but then lack of knowledge is no guarantee of success."

Despite this little broadside, Wolfgang had confirmed everything I myself had been thinking before I decided to make physics my career. I was very glad not to have tried my hand at pure mathematics, and I looked back on Lindemann's little dog as "part of that power which still produceth good, whilst ever scheming ill."

3

"Understanding"
in Modern Physics (1920–922)

My first two years at Munich University were spent in two quite different worlds: among my friends of the Youth Movement and in the abstract realm of theoretical physics. Both worlds were so filled with intense activity that I was often in a state of great agitation, the more so as I found it rather difficult to shuttle between the two. In Sommerfeld's seminar, talks with Wolfgang Pauli constituted the most important part of my studies. But Wolfgang's way of life was almost diametrically opposed to my own. While I loved the daylight and spent as much of my free time as I could mountain-walking, swimming or cooking simple meals on the shore of one of the Bavarian lakes, Wolfgang was a typical night bird. He preferred the town, liked to spend his evenings in some old bar or café, and would then work on his physics through much of the night with great concentration and success. Quite naturally, and to Sommerfeld's dismay, he would rarely attend morning lectures; it was not until noon that he generally turned up. This difference in our styles of living was the subject of quite a bit of ribbing, but did not otherwise mar our friendship—our common interest in physics saw to that.

When I think back on the summer of 1921, and try to compress my many memories into a single picture, my mind's eye conjures up a camp that stood on the edge of a forest. Below, still in the gray light of dawn, lay the lake in which we had swum the day before, and across it, in the distance, the broad crest of the Benedictine Ridge. My comrades would still be asleep when I left

my tent and took the footpath for an hour's walk to the nearest
railway station. From there, the early-morning train would carry
me to Munich in time to attend Sommerfeld's 9 A.M. lecture. The
footpath led down to the lake through marshy ground, then on to
a moraine with a view over the Alpine chain, from the Benedictine Ridge as far as the Zugspitze. On the blossoming meadows, I
could see the first mowing machines, and I was sorry that I was
no longer a farm worker on the Grossthalerhof in Miesbach.
Three years earlier, using a pair of oxen, I would have done my
best to cut the meadow so straight that no strip of grass, or, as the
farmer called it, no "pig," was left behind. And so, my thoughts
filled with a colorful medley of everyday impressions from my
farming days, with the beauty of the landscape, and with Sommerfeld's coming lecture, I was convinced I was the happiest
mortal on earth.

An hour or two after the end of Sommerfeld's lecture, Wolfgang would appear in the seminar, and our conversation would
go something like this:

Wolfgang: "Good morning. If it isn't our prophet of nature!
You look for all the world as if you have been living by the
principles of St. Jean-Jacques. Wasn't it he who said, 'Back to
nature, up into the trees, you apes'?"

"The second part of the quotation is not from Rousseau," I
would explain, "and none of us goes in for climbing trees. In any
case, it isn't morning; it's twelve o'clock, if my watch is correct.
One day you simply must introduce me to one of your nightly
haunts so that I, too, can get a whiff of your physical inspiration."

"That wouldn't help you in the least. Still, you might perhaps
care to tell me what you have managed to find out about
Kramers' work, on which you're supposed to lecture our class."

And so our talk would change quickly from good-humored
insults to more pressing matters. When we talked about physics,
we were often joined by our friend, Otto Laporte, whose sober,
pragmatic approach made him an excellent mediator between
Wolfgang and myself. He and Sommerfeld were later to publish
important papers on the multiplet structure of atomic spectra.

It was probably due to him that the three of us—Wolfgang,
Otto and I—decided to go on a bicycle tour that took us from
Benediktbauern across the Kesselberg to Lake Walchen and

on into the Loisach Valley. This was probably the only time Wolfgang dared to enter my world—with the most beneficial results for myself: the talks we began during that tour and continued in Munich were to have a lasting effect.

Meanwhile we spent a few happy days on the road. Once we had reached the saddle of the Kesselberg, laboring uphill with our bicycles, we could ride effortlessly along a road boldly cut into the mountain slope past the steep western shore of Lake Walchen—at the time I had no idea how important this little spot of earth was to become for me one day. We passed the very place where an old man and his daughter had once joined Goethe's coach en route to Italy, she the model of his future Mignon, he of the old harpist in *Wilhelm Meister!* Across this dark lake, Goethe caught his first glimpse of the snow-covered Alps. And while we, too, delighted in the glorious landscape, our conversation kept returning to our studies and to science in general.

Thus Wolfgang asked me—I think it was one evening at an inn in Cruimau—whether I at long last understood Einstein's relativity theory, on which Sommerfeld laid so much stress. I could only say that I did not really know what was meant by "understanding" in physics. The mathematical framework of relativity theory caused me no difficulties, but that did not necessarily mean that I had "understood" why a moving observer means something different by "time" than an observer at rest. The whole thing baffled me, and struck me as being quite "incomprehensible."

"But once you have grasped the mathematical framework," Wolfgang objected, "you can surely predict what an observer at rest and a moving observer ought to observe or measure. And we have good reason to assume that a real experiment will bear out these predictions. What more can you ask?"

"That is precisely my problem," I replied, "that I don't know what more I can ask. I feel somewhat cheated by the logic of the new mathematical framework. You might even say that I have grasped the theory with my brain, but not yet with my heart. I think that I don't have to study physics to know what 'time' is; after all, our every thought and action presupposes a naïve time concept. Perhaps I could put it like this: our thought depends on the fact that this time concept works, that we can operate with it. But if our naïve time concept has to be changed, then we can no longer tell whether our language and thought remain useful working tools. In saying this, I am not trying to hark back to Kant, for whom time and space were a priori forms of the intuition. In other words, to Kant, as to the earlier physicists, time and space were absolutes. I only want to stress the fact that our language and thought become vague whenever we try to change such basic concepts, and uncertainty goes ill with true understanding."

Otto found my scruples quite unnecessary.

"That's how it may look in the schoolmen's philosophy," he said, "but if they ascribed definite, immutable meanings to the concepts of time and space, that only goes to show that their philosophy was false. I can't do anything with beautiful phrases about the 'essence' of space and time. You've probably read more philosophy than is good for you. Remember this splendid definition: 'Philosophy is the systematic misuse of nomenclature specially invented for the purpose.' All absolute claims must be rejected a priori. We ought only to use such words and concepts as can be directly related to sense perception, with this proviso, of course, that we may substitute complex physical observations for direct perception. It is precisely this return to observable phenomena that is Einstein's great merit. In his relativity theory, he quite rightly started with the commonplace assumption that time is what you read off a clock. If you keep to this commonplace meaning, you will have few problems with relativity theory. As soon as a theory allows us to predict the results of an observation, it gives us all the understanding we need."

Wolfgang now brought up a number of objections: "What you say is true only under certain conditions, and these ought to be stated. To begin with, you have to be certain that your theoretical predictions are unambiguous and self-consistent. In the case of relativity theory, this is probably guaranteed by the simple mathematical framework. Next it must be quite clear from the conceptual structure of a theory to which particular phenomena it applies and to which it does not. In the absence of this qualification every theory can be refuted at once, simply because no theory can predict all the phenomena in the world. But even if all these conditions are met, I am still not altogether certain whether the ability to predict phenomena in a particular area entitles one to claim full understanding. Conversely, it may be quite possible to understand a particular realm of experience completely without being able to predict all the results of future observations."

I now tried to show that correct predictions were not necessarily a sign of true understanding by quoting an historical example. "You know, of course, that the Greek astronomer, Aristarchus, considered the possibility that the sun might occupy the center of our planetary system. This view was rejected by Hipparchus, and then fell into oblivion. Ptolemy started with the assumption that the earth was the central body, and he treated the orbits of the planets as superimposed cycles and epicycles. This enabled him to predict eclipses of the sun and the moon very precisely, so precisely that for fifteen hundred years his doctrine was considered the certain foundation of astronomy. But did Ptolemy really understand the planetary system? Was it not Newton who, knowing the law of inertia, and introducing force as the cause of changes of momentum, was the first to give a proper explanation of planetary motions in terms of gravitation? Was he not the first to have really understood this type of motion? This, to me, is a crucial question.

"Or let us take an example from the more recent history of physics. At the end of the eighteenth century, when electrical phenomena became better known, physicists were able to make very precise calculations of the electrostatic forces governing the behavior of charged bodies, treating them as centers of force in the manner of Newtonian mechanics. That much at least I gathered from Sommerfeld. But it was only when Faraday changed the entire problem and inquired into fields of force, i.e., into the distribution of forces in time and space, that he provided a true understanding of electromagnetic phenomena, and laid the foundations on which Maxwell could later base his mathematical formulae."

Otto did not find my examples particularly convincing. "I can see only differences in degree, but no basic distinction. Ptolemy's astronomy must have been very good, else it would not have lasted for fifteen hundred years. Newton's didn't seem much better at first; it took quite some time before astronomers came to appreciate that it led to more accurate predictions of the motions of the planets than Ptolemy's cycles and epicycles. I cannot really grant you that Newton did something fundamentally better than Ptolemy. He merely gave a different account of planetary motions, one that happened to prove more successful in the long run."

Wolfgang found this argument too one-sided and much too positivistic. "I, for one, see a basic distinction between Newton's astronomy and Ptolemy's," he said. "To begin with, Newton posed the whole problem quite differently: he inquired into the causes of planetary motions and not into the motions themselves. These causes, he discovered, were forces, and in our planetary system they happen to be much simpler than the motions. He described them by means of his law of gravitation. If we say that Newton helped us to understand the motion of the planets, we only mean that more precise observations have shown that it is possible to reduce the complicated motions of the planets to something very simple, namely, to gravitational forces, and to explain them in that way. Admittedly, Ptolemy could describe all the complicated motions of the planets by the superposition of cycles and epicycles, but he had to treat them as empirical facts. Moreover, Newton was also able to show that the motions of the planets are governed by the same laws as those that determine the motion of a projectile, the oscillation of a pendulum or the spinning of a top. The mere fact that Newton's mechanics reduced all these different phenomena to a simple principle, namely, 'mass $\times$ acceleration $=$ force,' shows that his planetary system is vastly superior to Ptolemy's."

Otto still refused to admit defeat. "The word 'cause' and the assertion that force is the cause of motion all sound very well, but in fact only take us a slight step forward. For we are then compelled to ask the next question: What are the causes of forces in general and of gravitation in particular? In other words, according to your own philosophy, we can only claim 'real' understanding of planetary motions once we know the cause of gravitation, and so on *ad infinitum*."

Wolfgang objected strongly to this argument. "Of course we can keep on asking questions," he said, "but isn't that the basis of all science? Your argument is not relevant to the point under discussion. 'Understanding' nature surely means taking a close look at its connections, being certain of its inner workings. Such knowledge cannot be gained by understanding an isolated phenomenon or a single group of phenomena, even if one discovers some order in them. It comes from the recognition that a wealth of experiential facts are interconnected and can therefore be reduced to a common principle. In that case, certainty rests precisely on this wealth of facts. The danger of making mistakes is the smaller, the richer and more complex the phenomena are, and the simpler is the common principle to which they can all be brought back. The fact that still wider connections may yet be discovered makes no difference at all."

"And so you think," I asked, "that we can trust in relativity theory on the grounds that it helps us to combine under a common heading, or reduce to a common root, a great wealth of phenomena, for instance in electrodynamics? If I understand you correctly, you are maintaining that, since in this case a uniform connection is readily established and can be shown to be mathematically transparent, we get the feeling that we have understood' relativity, even though we are forced to give the words 'space' and 'time' a new, or let us say a changed, meaning."

"Yes, I do mean something like that. The decisive steps of Newton and of Faraday were, in each case, their new way of asking questions and of formulating concepts by which the correct answers could be obtained. 'Understanding' probably means nothing more than having whatever ideas and concepts are needed to recognize that a great many different phenomena are part of a coherent whole. Our mind becomes less puzzled once we have recognized that a special, apparently confused situation is merely a special case of something wider, that as a result it can be formulated much more simply. The reduction of a colorful variety of phenomena to a general and simple principle, or, as the Greeks would have put it, the reduction of the many to the one, is precisely what we mean by 'understanding.' The ability to predict is often the consequence of understanding, of having the right concepts, but is not identical with understanding."

Otto murmured: " 'The systematic misuse of nomenclature

specially invented for the purpose.' I cannot for the life of me see why it is necessary to speak in such complicated ways about simple things. If we use language to refer to direct sense impressions, then few misunderstandings can arise—every word has a precise meaning, and if a theory sticks to that limitation, it will always be comprehensible, even without a lot of philosophizing."

But Wolfgang refused to accept this. "Your suggestion, which sounds so terribly plausible, has already been made, by Mach and others. It has even been said that Einstein arrived at his theory of relativity simply by sticking to Mach's doctrine. But this strikes me as a crude oversimplification. It is well known that Mach did not believe in the existence of atoms, on the grounds that they cannot be observed. For all that, atoms were needed to explain a host of physical and chemical phenomena that had eluded scientists in the past. Mach himself was obviously led astray by the very principle you defend, and, as far as I am concerned, this was not by chance."

"Everybody can make mistakes," Otto said, trying to calm us down. "But mistakes are no excuse for making things more complicated than they are. The theory of relativity is so simple that anyone can grasp it. But when it comes to atomic theory, things are very much more obscure."

This brought us to our second theme, which kept us busy long after our bicycle tour was over. It was to become the source of keen arguments in our Munich seminar, often in Sommerfeld's presence.

The central subject of Sommerfeld's seminar was Bohr's atomic theory. Basing his ideas on decisive experiments by Rutherford, Bohr had depicted the atom as a tiny planetary system with a central nucleus which, though considerably smaller than the atom, carried most of its mass. About this nucleus, a number of extremely lightweight electrons revolved like so many planets. However, while the orbits of planets were determined by known forces and the past history of the system, and hence subject to perturbations, the orbits of electrons were said to call for additional postulates of a special kind, postulates that helped to explain the peculiar stability of matter when exposed to external influences. Ever since Planck had published his famous work in 1900, these additional postulates were known as quantum conditions, and it was they which had introduced into atomic physics that strange element of number mysticism to which I referred earlier. Certain magnitudes that could be computed from an orbit were said to be integral multiples of a basic unit, namely, Planck's quantum of action. Such rules were highly reminiscent of Pythagorean ideas, according to which two vibrating strings were in harmony if, with equal tension, their lengths were in simple proportion. But what did the orbits of electrons have to do with vibrating strings? Even more confusing was the new explanation of light emission by atoms. In this process, a radiating electron was said to jump from one quantum orbit to the next, and to emit the energy thus liberated as a whole packet, or light quantum. Such ideas would never have been taken seriously had they not helped to explain a whole range of experiments with great accuracy.

This peculiar mixture of incomprehensible mumbo jumbo and empirical success quite naturally exerted a great fascination on us young students. Soon after the beginning of my studies, Sommerfeld set me a test: from the observations which an experimental physicist of his acquaintance had communicated to him, I was to deduce the electron orbits and quantum numbers involved. The task itself was not difficult, but the results proved extremely perplexing; apart from integral quantum numbers, I was also forced to admit halves, and this ran counter to the spirit of quantum theory and of Sommerfeld's number mysticism. Wolfgang suggested that I would soon have to introduce quarters and eighths as well, until finally the whole quantum theory would crumble to dust in my capable hands. And try as I might, I could not get rid of the embarrassing fraction.

Wolfgang had set himself a more difficult task. He wanted to find out whether in a more complicated system, one that could only be determined by astronomical computations, Bohr's theory and the Bohr-Sommerfeld quantum conditions would still lead to experimentally valid results. In fact, during our Munich discussions, some of us had begun to feel that the earlier successes of the theory might have been due to the use of particularly simple systems, and that the theory would break down in a slightly more complicated one.

In this connection, Wolfgang asked me one day: "Do you

honestly believe that such things as electron orbits really exist inside the atom?"

My answer may have sounded a little labored. "To begin with," I told him, "we can observe the path of an electron in a cloud chamber: it leaves a clear trail of fog where it has passed. And since there is such a thing as an electron trajectory in the cloud chamber, we may take it that it will occur in the atom as well. But I have some reservations on that score. For while we determine the path itself by classical Newtonian methods, we use quantum conditions to account for its stability, thus flying in the face of Newtonian mechanics. And when it comes to electrons jumping from one orbit into the next—as the theory demands— we are careful not to specify whether they make high jumps, long jumps or some other sorts of jump. It all makes me think that something is radically wrong with the whole idea of electron orbits. But what is the alternative?"

Wolfgang nodded. "The whole thing seems a myth. If there really were such a thing as an electron orbit, the electron would obviously have to revolve periodically, with a given frequency. Now, we know from electrodynamics that if an electrical charge is set in periodic motion, it must emit electrical vibrations, i.e., radiate light of a characteristic frequency. But this is not supposed to happen with the electron; instead, the frequency of vibration of the emitted light is said to lie somewhere between the orbital frequency before the mysterious jump and the orbital frequency after the jump. All this is sheer madness."

"'Though this be madness, yet there is method in 't,'" I quoted.

"Yes, perhaps. Niels Bohr claims that he can tell the electron orbits of every atom in the periodic system, and the two of us do not even believe in the existence of such orbits. Sommerfeld perhaps disagrees with us. And, in fact, anyone can see electronic orbits in a cloud chamber. Quite likely Niels Bohr is right in a sense, though we cannot tell precisely in what sense."

Unlike Wolfgang, I felt optimistic about the issue, and I may have said something like this: "I find Bohr's physics most fascinating, difficulties and all. Bohr must surely know that he starts from contradictory assumptions which cannot be correct in their present form. But he has an unerring instinct for using these very assumptions to construct fairly convincing models of atomic processes. Bohr uses classical mechanics or quantum theory just as a painter uses his brushes and colors. Brushes do not determine the picture, and color is never the full reality; but if he keeps the picture before his mind's eye, the artist can use his brush to convey, however inadequately, his own mental picture to others. Bohr knows precisely how atoms behave during light emission, in chemical processes and in many other phenomena, and this has helped him to form an intuitive picture of the structure of different atoms; a picture he can only convey to other physicists by such inadequate means as electron orbits and quantum conditions. It is not at all certain that Bohr himself believes that electrons revolve inside the atom. But he is convinced of the correctness of his picture. The fact that he cannot yet express it by adequate linguistic or mathematical techniques is no disaster. On the contrary, it is a great challenge."

Wolfgang remained skeptical. "I must first of all find out whether the Bohr-Sommerfeld assumptions will lead to reasonable results in the case of my problem. If not—and I almost suspect that I shall discover just that—I shall at least know what does *not* work, and that, too, is a great step forward." Then he added reflectively: "Bohr's pictures may be right after all. But what are we to make of them, and what laws do they express?"

Some time later, Sommerfeld asked me rather unexpectedly, after a long talk about atomic theory: "Would you like to meet Niels Bohr? He is about to give a series of lectures in Göttingen. I have been invited, and I should like to take you along." I hesitated for a moment—the fare to Göttingen and return was quite beyond my financial resources. Perhaps Sommerfeld saw the shadow flit across my face. In any case, he quickly added that he would see to my expenses, whereupon I accepted with gratitude and alacrity.

In the early summer of 1922 Göttingen, that friendly little town of villas and gardens on the slopes of the Hain Mountain, was a mass of blooming shrubs, rose gardens and flower beds. Nature herself seemed to approve the name we later gave those wonderful days: the Göttingen Bohr Festival. I shall never forget the first lecture. The hall was filled to capacity. The great Danish physicist, whose very stature proclaimed him a Scandinavian, stood on the platform, his head slightly inclined, and a friendly but somewhat embarrassed smile on his lips. Summer light flooded in through the wide-open windows. Bohr spoke fairly softly, with a slight Danish accent. When he explained the individual assumptions of his theory, he chose his words very carefully, much more carefully than Sommerfeld usually did. And each one of his carefully formulated sentences revealed a long chain of underlying thoughts, of philosophical reflections, hinted at but never fully expressed. I found this approach highly exciting; what he said seemed both new and not quite new at the same time. We had all of us learned Bohr's theory from Sommerfeld, and knew what it was about, but it all sounded quite different from Bohr's own lips. We could clearly sense that he had reached his results not so much by calculation and demonstration as by intuition and inspiration, and that he found it difficult to justify his findings before Göttingen's famous school of mathematics. Each lecture was followed by long discussions, and at the end of the third lecture I myself dared to make a critical remark.

Bohr had been talking about Kramers' contribution—the subject on which I had been asked to speak in Sommerfeld's seminar —and he concluded that, although the basis of Kramers' theory was still unexplained, it seemed certain that the results were correct and would one day be confirmed by experiment. I then rose and advanced objections to Kramers' theory based on our Munich discussions.

Bohr must have gathered that my remarks sprang from profound interest in his atomic theory. He replied hesitantly, as though he were slightly worried by my objection, and at the end of the discussion he came over to me and asked me to join him that afternoon on a walk over the Hain Mountain. There we might go more deeply into the whole problem.

This walk was to have profound repercussions on my scientific career, or perhaps it is more correct to say that my real scientific career only began that afternoon. A well-tended mountain path took us past a popular coffeehouse, Zum Rohns, to a sunlit height, from which we looked down on the small university town, dominated by the spires of the old churches of St. John and St. Jacob and, beyond, across the Leine Valley.

Bohr opened the conversation. "This morning," he said, "you expressed some reservations about Kramers' work. I must tell you at once that I fully understand the nature of your doubts. Perhaps I ought to explain where I stand myself. Basically, I agree with you much more than you might think; I realize full well how cautious one has to be with assertions about the structure of atoms. I had best begin by telling you a little about the history of this theory. My starting point was not at all the idea that an atom is a small-scale planetary system and as such governed by the laws of astronomy. I never took things as literally as that. My starting point was rather the stability of matter, a pure miracle when considered from the standpoint of classical physics.

"By 'stability' I mean that the same substances always have the same properties, that the same crystals recur, the same chemical compounds, etc. In other words, even after a host of changes due to external influences, an iron atom will always remain an iron atom, with exactly the same properties as before. This cannot be explained by the principles of classical mechanics, certainly not if the atom resembles a planetary system. Nature clearly has a tendency to produce certain forms—I use the word 'forms' in the most general sense—and to recreate these forms even when they are disturbed or destroyed. You may even think of biology: the stability of living organisms, the propagation of the most complicated forms which, after all, can exist only in their entirety. But in biology we are dealing with highly complex structures, subject to characteristic, temporary transformations of a kind that need not detain us here. Let us rather stick to the simpler forms we study in physics and chemistry. The existence of uniform substances, of solid bodies, depends on the stability of atoms; that is precisely why an electron tube filled with a certain gas will always emit light of the same color, a spectrum with exactly the same lines. All this, far from being self-evident, is quite inexplicable in terms of the basic principle of Newtonian physics, according to which all effects have precisely determined causes, and according to which the present state of a phenomenon or process is fully determined by the one that immediately preceded it. This fact used to disturb me a great deal when I first began to look into atomic physics.

"The miracle of the stability of matter might have gone unnoticed even longer had experiments during the past few decades not thrown fresh light on the whole subject. Planck, as you know, discovered that the energy of an atomic system changes discontinuously; that when such a system emits energy, it passes through certain states with selected energy values. I myself later coined the term 'stationary states' for them. Next came Rutherford's crucial studies of the structure of the atom. It was in Rutherford's Manchester laboratory that I first became acquainted with the problems involved. At the time, I was barely older than you are today, and I kept plying Rutherford with long questions. Physicists had just begun to take a closer look at luminous phenomena and were busily determining the characteristic spectral lines of the various chemical elements; needless to say, chemists, too, produced a wealth of information on the behavior of atoms. These developments, which I was privileged to witness at close quarters, naturally made me wonder how all these things hung together. The theory I tried to put forward was meant to do no more than establish that connection.

"Now, this was really a hopeless task, quite different from those physicists normally tackle. For in all previous physics, or in any other branch of science, you could always try to explain a new phenomenon by reducing it to known phenomena or laws. In atomic physics, however, all previous concepts have proved inadequate. We know from the stability of matter that Newtonian physics does not apply to the interior of the atom; at best it can occasionally offer us a guideline. It follows that there can be no descriptive account of the structure of the atom; all such accounts must necessarily be based on classical concepts which, as we saw, no longer apply. You see that anyone trying to develop such a theory is really trying the impossible. For we intend to say something about the structure of the atom but lack a language in which we can make ourselves understood. We are in much the same position as a sailor, marooned on a remote island where conditions differ radically from anything he has ever known and where, to make things worse, the natives speak a completely alien tongue. He simply must make himself understood, but has no means of doing so. In that sort of situation a theory cannot 'explain' anything in the usual strict scientific sense of the word. All it can hope to do is to reveal connections and, for the rest, leave us to grope as best we can. That is precisely what Kramers' calculations were intended to do; perhaps I failed to stress this sufficiently at my lecture. And to do more than that is quite beyond our present means."

From Bohr's remarks it was quite obvious that he was familiar with all the doubts we ourselves had been expressing. But to make doubly sure that I had understood him, I asked: "If that is all we can do, what is the point of all those atomic models you produced and justified during the past few lectures? What exactly did you try to prove with them?"

"These models," Bohr replied, "have been deduced, or if you prefer guessed, from experiments, not from theoretical calculations. I hope that they describe the structure of the atoms as well, but *only* as well, as is possible in the descriptive language of classical physics. We must be clear that, when it comes to atoms, language can be used only as in poetry. The poet, too, is not nearly so concerned with describing facts as with creating images and establishing mental connections."

"But in that case how are we ever to make progress? After all, physics is supposed to be an exact science."

"It seems likely that the paradoxes of quantum theory, those incomprehensible features reflecting the stability of matter, will become sharper with every new experiment. If that happens, we can only hope that, in due course, new concepts will emerge which may somehow help us to grasp these inexpressible processes in the atom. But we are still a long way from that."

Bohr's remark reminded me of Robert's comment, during our walk near Lake Starnberg, that atoms were not things. For although Bohr believed that he knew a great many details about the inner structure of atoms, he did not look upon the electrons in the atomic shell as "things," in any case not as things in the sense of classical physics, which worked with such concepts as position, velocity, energy and extension. I therefore asked him: "If the inner structure of the atom is as closed to descriptive accounts as you say, if we really lack a language for dealing with it, how can we ever hope to understand atoms?"

Bohr hesitated for a moment, and then said: "I think we may yet be able to do so. But in the process we may have to learn what the word 'understanding' really means."

Our little walk had taken us to the peak of the Hain Mountain, to the famous Kehr Inn, so called because since olden times people used to turn back here [*umkehren*]. We, too, now made for the lowland, this time in a southerly direction, and looked down over the hills, woods and villages of the Leine Valley, long since incorporated into Göttingen town.

"We have talked about so many difficult subjects," Bohr continued, "and I have told you how I myself first got into this whole business; but I know nothing at all about you. You look very young. From your questions, it almost seems as if you started with atomic theory first, and then went on to take a look at orthodox physics. Sommerfeld must have introduced you to this adventurous world of atoms at a very early age. Do tell me about it and also about what you did in the war."

I confessed that, being twenty, I was only in my fourth term at the university, and that I knew very little indeed about general physics. I went on to tell him about Sommerfeld's class, where I had been especially attracted by the mysterious, inexplicable features of quantum theory. I added that I had been too young to serve in the army, but that my father had fought in France as a reserve officer and I had been very anxious about him. He was wounded in 1916 and was sent back home. During the last year of the war, I worked as a farm laborer in the Lower Bavarian Alps, to keep body and soul together. Otherwise I had been spared by the war.

"I should like to hear a lot more from you," Bohr said, "and to learn more about conditions in your country, of which I know so little. And about the Youth Movement, of which I have heard so much from my colleagues in Göttingen. You must pay us a visit in Copenhagen; perhaps you could stay with us for a term, and we might do some physics together. And then I'll show you round our small country and tell you about its history."

As we approached the edge of the town, the conversation turned to Göttingen's leading physicists and mathematicians— Max Born, James Franck, Richard Courant and David Hilbert, all of whom I had only just met. Bohr suggested that I might do part of my studies under them. Suddenly the future looked full of hope and new possibilities, which, after seeing Bohr home, I painted to myself in the most glorious colors all the way back to my lodgings.

4

Lessons in Politics
and History (1922–1924)

The summer of 1922 ended on what, for me, was a rather saddening note. My teacher, Sommerfeld, had suggested that I attend the Congress of German Scientists and Physicians in Leipzig, where Einstein, one of the chief speakers, would lecture on the general theory of relativity. My father had bought me the return-trip ticket from Munich, and I was looking forward greatly to this chance of hearing the discoverer of relativity theory in person. Once in Leipzig, I moved into one of the cheapest inns in the poorest quarter of the city—I could afford nothing better. Then I made for the meeting hall, where I found a number of the younger physicists whose acquaintance I had made in Göttingen during the "Bohr Festival," and asked them about Einstein's lecture, scheduled within a few hours. I noticed a certain tension all around me, which struck me as being rather odd, but then Leipzig was not Göttingen. I filled in the waiting time with a walk to the Memorial (to the great Battle of Leipzig), where, hungry and exhausted by the overnight railway journey, I lay down on the grass and at once fell asleep. I was wakened by a young girl who had decided to pelt me with plums. She sat down beside me, and made her peace with me with generous offerings of fruit from her ample basket.

The lecture theater was a large hall with doors on all sides. As I was about to enter, a young man—I learned later that he was an assistant or pupil of a well-known professor of physics in a South German university—pressed a red handbill into my hand, warning me against Einstein and relativity. The whole theory was said to be nothing but wild speculation, blown up by the Jewish press and entirely alien to the German spirit. At first I thought the whole thing was the work of some lunatic, for madmen are wont to turn up at all big meetings. However, when I was told that the author was a man renowned for his experimental work, to whom Sommerfeld had often referred in his lectures, I felt as if part of my world were collapsing. All along, I had been firmly convinced that science at least was above the kind of political strife that had led to the civil war in Munich, and of which I wished to have no further part. And now I made the sad discovery that men of weak or pathological character can inject their twisted political passions even into scientific life. Needless to say, my immediate reaction was to drop any reservations I may have had with regard to Einstein's theory, or rather to what I knew about it from Wolfgang's occasional explanations. For if I had learned one thing from my experiences during the civil war, it was that one must never judge a political movement by the aims it so loudly proclaims and perhaps genuinely strives to attain, but only by the means it uses to achieve them. The choice of bad means simply proves that those responsible have lost faith in the persuasive force of their original arguments. In this instance, the means applied by a leading physicist in his attempt to refute the theory of relativity were so bad and insubstantial that they could signify only one thing: the man had abandoned all hope of ever refuting the theory with scientific arguments.

Still, so upset was I by this spectacle that I failed to pay proper attention to Einstein himself, and, at the end of the lecture, forgot to avail myself of Sommerfeld's offer to introduce me to the speaker. Instead, I returned somberly to my inn, only to discover that all my possessions—rucksack, linen, socks and second suit—had been stolen. Luckily I still had my return ticket. I went to the station and took the next train to Munich. I was in utter despair because I knew that my father would find it extremely hard to make up my loss. And so, upon discovering that my parents were out of the city, I took a job as a woodman in Forstenried Park, south of the town. The pines there had been attacked by bark beetles, and a large number of trees had to be felled and their bark burned. Only when I had earned enough money to replenish my meager wardrobe did I return to my studies.

I have mentioned this whole unhappy episode, not to resurrect events that are best forgotten, but because it later cropped up in my conversations with Niels Bohr and affected my behavior in the dangerous no-man's land between science and politics. At first, the Leipzig experience left me with a deep sense of disappointment and with doubts about the validity of science in general. For if science, too, was more concerned with private feuds than with discovering the truth, was it really worth bothering with? Luckily, in the end, memories of my walk with Niels Bohr prevailed over all such pessimistic thoughts, and I was hopeful that I might one day avail myself of Bohr's generous invitation and have many more talks with him in Copenhagen.

As it happened, a year and a half were to go by before this came to pass. Meanwhile I spent a term at Göttingen, submitted a thesis on the stability of laminar flow in fluids, sat for my examination in Munich and, for another term, served as Max Born's assistant in Göttingen. During the Easter vacation of 1924 I finally boarded the Warnemünde ferry for Denmark. Throughout the trip I feasted my eyes on a host of colorful boats, including four-masters in full rig. At the end of the First World War, a large part of the world's merchant fleet had ended up at the bottom of the sea, with the result that the old sailing boats had to be brought out again, and the seascape looked all the brighter for it—much as it had done a hundred years before. When I eventually disembarked, I had some trouble with customs—I knew no Danish and could not account for myself properly. However, as soon as it became clear that I was about to work in Professor Bohr's Institute, all difficulties were swept out of the way and all doors were opened to me. And so from the very outset I felt safe under the protection of one of the greatest personalities in this small but friendly country.

Not that my first few days in the Institute were particularly easy for me. I suddenly came face to face with a large number of brilliant young men from every part of the world, all of them greatly superior to me, not only in linguistic prowess and worldliness, but also in their knowledge of physics. I saw very little of Bohr himself; he obviously had his hands full with administrative tasks, and I obviously had no right to make greater claims on his time than had other members of the Institute. But after a few days, he came into my room and asked me to join him for a few days' walking tour through the island of Zealand. In the Institute itself, he said, there was little chance for lengthy talks, and he wanted to get to know me better.

And so the two of us set out with our rucksacks. First we took the trolley to the northern edge of the city, and from there we walked through the Deer Park, once a hunting preserve. We admired the beautiful little Hermitage Castle right in its center, and watched large herds of deer graze in the clearings. Then we made for the north, sometimes hugging the coast, sometimes walking through forests and past peaceful lakes, studded with summer houses still sleeping behind closed shutters—it was early spring and the trees were only just putting out their first tender shoots. Our talk turned to conditions in Germany, and Bohr asked what I remembered about the outbreak of war, ten years before.

"I have heard a great deal about those days," he told me. "Friends of ours who traveled through Germany early in August 1914 spoke of a great wave of enthusiasm that gripped not only the whole German nation but even outsiders, whose emotions, however, were tempered with horror. Isn't it odd that a whole people should have gone into war in a flush of war fever, when they ought to have known how many friends and enemies alike that war would swallow up, how many injustices would be committed by both sides? Can you explain any of this?"

"I was only twelve at the time," I replied, "and obviously my opinions were based on what I picked up from conversations between my parents and grandparents. Still, I don't think the words 'war fever' quite describe the situation. No one I knew was happy about what lay before us, and no one was pleased that the war had started. If you ask me to describe what happened, I would say we suddenly realized that things had become serious. We felt that we had all been living in a world of dreams, and that this beautiful world had suddenly been shattered by the murder of the heir to the Austrian crown. Suddenly we were face to face with reality, with a call none of us could refuse, a call we had to answer come what may—with heavy hearts, but with all our hearts nonetheless. Needless to say, we were all convinced of the justice of the German cause, for Germany and Austria were like one country, and the murder of the Archduke Francis Ferdinand and his wife by members of a secret Serbian society struck us as a crime against us all. So we had to defend ourselves, and, as I have said, most Germans decided to do so wholeheartedly.

"Now, such popular decisions have something highly seductive about them, something quite uncanny and irrational—even I could feel that in August 1914. I was traveling with my parents from Munich to Osnabrück, where my father, who was a captain in the Reserves, had to report for duty. All the railway stations were filled with shouting crowds of excited people; freight cars, decorated with flowers and branches, were packed with soldiers and guns. Young women and children stood alongside them; there was much crying and singing until the train left the station. You could address any stranger you wanted to as if you were old friends; everyone helped everyone else—we had all become brothers in fate. I should not like to eradicate this day from my memory. And yet this incredible, this unimaginable day, a day no one who witnessed it could forget, had nothing to do with what is commonly called war fever. I think that the whole thing was distorted after the event."

"You must realize," Bohr said, "that we, in our small country, take quite a different view of such matters. Look at it historically. Perhaps Germany's expansion during the last century proved just a little too easy. There was first of all the war against our country in 1864, which caused so much bitterness among us, then your victory over Austria in 1866 and over France in 1870. To Germans, it must have looked as if a great Central European Empire could be built almost overnight. But things aren't ever that simple. To found empires one must first win the hearts of the people. This the Prussians, for all their efficiency, obviously failed to do; perhaps because their way of life was too hard, or perhaps because their ideas of discipline did not appeal to others. By the time Germany came to realize that, it was too late. In any case, the German attack on the small country of Belgium struck the outside world as an act of blatant aggression, in no way justified even by the assassination of the Austrian heir. After all, Belgium had nothing to do with the assassination, nor was it a party to any alliance against Germany."

"Certainly we Germans committed a great many wrongs in

that war," I had to admit, "just as our opponents did. But then war is bound to lead to wrongs. And I must also admit that the only tribunal competent to decide the issue—world history—has found against us. Otherwise, I am probably much too young to judge which politicians made the right or wrong decisions in which places. But there are two things that have always bothered me, and I should like to know what you think about them.

"I told you that when war was declared, the whole world seemed completely changed. All petty, everyday cares suddenly disappeared. Personal and family relationships, once the very center of our lives, gave way to the broader solidarity of a whole nation sharing a common fate. Houses, streets, forests—everything looked quite different, or, as Jakob Burckhardt has put it, 'Heaven itself took on a fresh hue.' My best friend, a cousin from Osnabrück, who was a few years older than I, became a soldier. I do not know whether he was conscripted or whether he volunteered. Such questions were never even asked. The great decision had been made, everyone who was physically fit joined the army. My cousin would never have had the least wish to make war on anyone, or to fight for German conquests, though he was certain of our victory. So much I gathered from our last conversation just before he left. All we knew was that he was expected to offer his life, like all the rest. He may for a moment have been deeply frightened, but still he said yes like everyone else. Had I been a few years older, I would probably have done the same thing. My cousin died in France. Do you think he ought to have told himself that the whole war was nonsense, a fever, mass suggestion, and have refused this call on his life? Who has the right to decide? A young man who could not possibly hope to see through the machinations of world politics, who knew no more than a few facts, difficult enough to grasp in themselves: a murder in Sarajevo or our invasion of Belgium?"

"What you tell me makes me very sad," Bohr replied, "for I think I can see what you are getting at. Perhaps what these young men felt as they went to war, certain of their cause, is part of the greatest happiness men can experience. But surely that is a terrible truth. When men go to war, don't you feel that they somehow resemble migratory birds who flock together in the autumn before heading south? None of these birds knows which has decided on the flight or even why they must migrate, but each is gripped by the same prevailing agitation, the same wish to join in, even if the flight leads to death. In human beings, the remarkable fact is that, on the one hand, the reaction is as elemental and as uncontrollable as, for instance, a forest fire or any other natural phenomenon, while, on the other hand, it releases a sense of almost boundless individual freedom. The young man who goes to war has thrown off the burden of his daily cares and worries. When life or death is at stake, petty reservations, all those qualms that normally restrict our lives, are cast to the winds. We have only one aim—victory—and life seems simple and clear as never before. There is probably no more beautiful description of this unique situation in the life of a young man than the trooper's song in Schiller's *Wallenstein*. You must know the last lines: 'Who would share life must risk it, and none who refuse the hazard shall gain it—who risks it may lose!' This is probably quite true. Yet for all that, we must say no, must make every effort to avoid wars, indeed all international conflicts from which wars arise. Our walk through Denmark may be a small step in that direction."

"I should like to put my second question, if I may," I continued. "You have spoken of the Prussian sense of discipline, and have told me that it does not appeal to other people. I myself grew up in southern Germany, and our tradition is such that we think somewhat differently from people born between Magdeburg and Königsberg. Yet the principles of Prussian life—the subordination of individual ambition to the common cause, modesty in private life, honesty and incorruptibility, gallantry and punctuality—have always attracted me. Even though these principles have been misused by politicians, I cannot really despise them. Why do you Danes, for instance, feel so differently?"

"I believe," Bohr said, "that we do appreciate the virtues of this Prussian attitude. But we prefer to give greater scope to the individual, to his dreams and plans, than the Prussian principle permits. We wish to be part of a community of free people, each of whom fully recognizes the rights of all the others. Freedom and individual independence are more important to us than strength derived from external discipline. It is very strange, isn't it, that our ideas of the good life should so often be molded by historical models, which have survived only in myths or legends, and yet retain their hold on us. The Prussian, I believe, models himself on the Teutonic knight, who swore the monk's vow of poverty, chastity and obedience, who spread the Christian light, sword in hand. We in Denmark prefer the heroes of the Icelandic sagas, the poet Egill, son of Skallagrim, who at the tender age of three defied his father, fetched himself a horse and followed Skallagrim on his long ride. Or the wise Njáll, who was better versed in the law than all men on the island, and whose advice was sought in all disputes. These men, or their ancestors, had gone to Iceland because they did not want to bend to the will of the mighty Norwegian kings. They refused to serve masters who could order them into a war that was the king's and not their own. They were all of them brave warriors and I am afraid lived chiefly on piracy. When you read these sagas, you will probably be horrified.by all the talk of fighting and killing. But these men wanted above all to be free, and they respected the right of others to be as free as they were themselves. They fought over possessions or honor, but not for power over others.

"Naturally, we cannot tell to what extent these sagas are based on historical fact. But within these terse chronicles of life in Iceland we can sense a great poetic force, so that it is not surprising that they should have continued to mold our ideas of freedom to this day. Life in Britain, also, where the Normans were once so prominent, has been stamped by this spirit of independence. The British form of democracy, the Englishman's sense of fairness and respect for the ideas and interests of others, his high regard for justice and law, may well derive from the same source. No doubt, that is why the British were able to build up a great empire. Admittedly, they, too, were guilty of acts of violence, much as the old Vikings were."

It was afternoon now, and we were walking close to the shore, through small fishermen's villages. Across the Öresund we could see the Swedish coast a few miles away bathed in the setting sun. When we reached Helsingör, it was getting dark, but we decided to take a quick walk through the precincts of Kronborg Castle, which dominates the narrowest part of the Öresund, and whose ramparts still bristle with old guns, symbols of a power long since gone. Bohr began to tell me about the history of this castle. Frederick II of Denmark had it built toward the end of the sixteenth century, in the Dutch Renaissance style. The walls no less than the bastion jutting out far into the Öresund serve as reminders of its military past. In the seventeenth century Swedish prisoners of war were still locked up in its casements. But now, as we stood next to the old guns in the dusk looking alternately across at the sailing boats on the Öresund and the tall Renaissance building behind us, we clearly sensed the harmony of a spot in which struggle had long since ceased. True, you still feel the pull of forces that once drove men against one another, destroying ships, raising cries of victories and screams of despair, but you also know that they no longer shape people's lives. One gets a direct, almost physical sense of peace all around.

Kronborg Castle, or rather the spot on which it stands, is connected with the legend of Hamlet, the Danish Prince who went mad or shammed madness to escape the machinations of his murderous uncle. Bohr mentioned the legend and went on to say: "Isn't it strange how this castle changes as soon as one imagines that Hamlet lived here? As scientists we believe that a castle consists only of stones, and admire the way the architect put them together. The stones, the green roof with its patina, the wood carvings in the church, constitute the whole castle. None of this should be changed by the fact that Hamlet lived here, and yet it is changed completely. Suddenly the walls and the ramparts speak a quite different language. The courtyard becomes an entire world, a dark corner reminds us of the darkness in the human soul, we hear Hamlet's 'To be or not to be.' Yet all we really know about Hamlet is that his name appears in a thirteenth-century chronicle. No one can prove that he really lived, let alone that he lived here. But everyone knows the questions Shakespeare had him ask, the human depths he was made to reveal, and so, he, too, had to be found a place on earth, here in Kronborg. And once we know that, Kronborg becomes quite a different castle for us."

While we were talking, dusk had turned almost into night; a cold wind was blowing across the Öresund and forced us to leave.

By next morning, the wind had freshened considerably. The

sky was swept clean, and across the bright blue Baltic we could see the Swedish coast as far as Kullen Peninsula in the north. We walked westward along the northern shore of Zealand, some seventy to a hundred feet above sea level and here and there sheer above the waves. Looking across to Kullen, Bohr said: "You grew up in Munich close to the mountains, and you have told me a great deal about your mountain walks. I know that mountain-dwellers must find Denmark terribly flat and boring. Perhaps you will never be able to like my country. But to us the sea is all-important. As we look across it, we think that part of infinity lies within our grasp."

"I can sense that," I replied, "and I noticed it particularly in the faces of the fishermen we met yesterday on the beach—people here have a distant, serene look. In the mountains things are quite different. There the eye passes from the nearby detail over rather complicated rock formations or icy peaks straight up to the sky. Perhaps that is why our people are so gay."

"We have only one mountain in Denmark," Bohr explained. "It is just over five hundred feet high and strikes us as so mag-nificent that we call it the Heavenly Peak. It is said that, when one of our compatriots tried to impress a Norwegian friend with this splendid phenomenon, the visitor looked at it disdain-fully and said, 'That's what we call a dump in Norway.' I hope you won't be so hard on our landscape. But please tell me some-thing about your own mountain walks with friends from the Youth Movement."

"We often set out for several weeks at a time. Last summer we went from Würzburg across the Rhön Mountains as far as the southern edge of the Harz Mountains, and from there by way of Jena and Weimar back into the Thuringian Forest and on to Bamberg. When it's warm enough, we usually sleep out in the open, but more often we sleep in a tent, or, if the weather is too bad, in a farmer's hay barn. Sometimes to pay for our shelter, we help with the harvest, and occasionally, if we make ourselves particularly useful, we get all sorts of wonderful farm fare as well. Otherwise we cook for ourselves, generally over a campfire in the forest, and in the evenings we read stories by the light of the logs or we sing or play music. Members of the Youth Movement have collected many old folk songs and have arranged them for parts with violin and flute accompaniment. Such music-making gives us all a great deal of pleasure, even though we often play it badly rather than well.

"Perhaps we sometimes imagine ourselves in the medieval role of traveling scholars, and compare the catastrophe of the last war and the subsequent political strife with the hopeless confusion of the Thirty Years' War, which in spite of its horrors is said to have inspired many of these songs. A feeling of kinship with that age seems to have seized young people all over Germany. I remember being stopped in the street by an unknown boy, who asked me to join a mass meeting of young people in an ancient castle. And, indeed, when I got there, scores of young people were already streaming toward the place, which stands in a most picturesque spot in the Swabian Jura and looks down from an almost vertical rock into the Altmühl Valley. I was quite overcome by the forces generated at this spontaneous gathering, much as I was on the first of August, 1914. Otherwise, our Youth Movement has very little to do with political issues."

"The life you describe seems highly romantic, and I would be quite tempted to share it. Moreover, I can see that you are swayed by the chivalrous ideals of which we spoke yesterday. But you don't have to take an oath, do you, before you join, as the Freemasons do?"

"No, there is no written or even unwritten rule to which we have to adhere. Most of us are far too skeptical for any such rituals. But perhaps I ought to add that we do observe certain rules, although no one orders us to. For instance, we don't smoke, we drink very little, we dress far too simply for our parents' liking, and I don't think any of us are very interested in night life or in bars—but there is no written code."

"And what happens if one of you breaks these rules?"

"I don't know, perhaps we simply laugh at him. But it just doesn't happen."

"Isn't it uncanny, or perhaps I should say marvelous," said Bohr, "how much magical power the old images retain? That after so many centuries they should still affect people, without written laws or external coercion? We spoke yesterday of monastic vows, and the monk's first two rules are highly commendable. Nowadays they amount to modesty and a willingness to adopt a somewhat harder, more continent life. But I hope you won't stress the third rule, obedience, too soon, or else there may be dangerous political consequences. You know that I think far more highly of the two Icelanders, Egill and Njáll, than of the masters of your Prussian orders.

"But you have told me that you were present during the civil war in Munich. You must surely have wondered about such general questions as the role of the state in the life of society. What bearing does all this have on your life in the Youth Movement?"

"During the civil war," I replied, "I sided with the government because the whole fight seemed quite senseless to me, and I hoped that it would end more quickly that way. But I must admit that I had a rather bad conscience toward our opponents. Ordinary Germans, and particularly the workers, had fought wholeheartedly for our victory in the war, had made the same sacrifices as everyone else. Their criticism of the ruling classes was absolutely justified, for our rulers had confronted the German people with an insoluble problem. Hence I felt that it was absolutely essential to make friendly contacts with the workers as soon as the civil war was over. That was also the view of a great many members of the Youth Movement.

"Four years ago, for instance, we helped run extracurricular classes in Munich, and I myself was rash enough to give a series of lectures on astronomy, pointing out the various constellations to some hundreds of workers and their wives, describing the motions of the planets and their distances from each other, and trying to interest them in the structure of our Milky Way. With a young lady, I also helped give a course of lectures on the German opera. She sang arias and I accompanied her on the piano; and afterward she would give brief summaries of the history and structure of the various operas. The whole thing was amateurish in the extreme, but I do believe that the audience appreciated our good intentions and that they enjoyed our recitals as much as we did. This was also the time when many young people in the Youth Movement turned to elementary school teaching, as a result of which I imagine that many of our elementary schools have much better teachers than quite a few of our so-called high schools.

"I quite understand why people abroad might look upon our Youth Movement as too romantic and idealistic, and why they are afraid it might be diverted into the wrong political channels. But I have no fears on that score, certainly not in the immediate present—after all, a great deal of good has already come out of the movement. I am thinking particularly of the revival of interest in old music—in Bach, plain song and ballads—of attempts to revive the old handicrafts, to bring beauty into the homes of even the very poor, and of all the efforts to awaken interest in the arts through amateur dramatic or music groups."

"I'm glad to see that you're so optimistic," Bohr said. "But now and then our papers also tell us about more ominous, anti-Semitic, trends in Germany, obviously fostered by demagogues. Have you come across any of that yourself?"

"Yes, in Munich such groups have begun to make quite a bit of a noise. They enjoy the support of some of the old officers who've been unable to come to terms with Germany's defeat. But we don't take these groups very seriously. After all, you can't base rational politics on resentment alone. What is far worse is that reputable scientists should see fit to repeat all this stuff like so many parrots."

And I told him of my experiences in Leipzig, where relativity theory had been the subject of political slanders. At the time neither of us had the least idea just what terrible consequences would one day spring from these apparently unimportant political aberrations, but more of this later. At the time Bohr's reply was directed at the resentful old officers as much as at the physicist who refused to come to terms with relativity theory.

"You see, once again I prefer the English attitude to the German. The English try to play the game, but they also try to be good losers. Prussians, on the other hand, think that losing is a disgrace, though they, too, preach magnanimity in victory, and that I find highly creditable. But the English go one step further: they expect the vanquished to be magnanimous to the victor, to accept their defeat and to bear no grudges. If they can, they have achieved the next best thing to victory. They are free men among free people. I am reminded of the old Vikings again.

Perhaps that makes me a romantic in your eyes, but I take the whole thing much more seriously than you may perhaps believe."

"Oh, no, I see how serious you are," I told him.

We had meanwhile reached Gilleleje, on the northern tip of Zealand. The beach, which in summer is crowded with happy holiday makers, was utterly deserted on this cold day. We picked up a few flat stones and tried our skill at making them skim the water, or aimed them at old fishermen's baskets or bits of driftwood. Bohr told me that, shortly after the war, he had visited this beach with Kramers, and that they had spotted a German mine with its detonator sticking up above the waves. They had tried to throw stones at the detonator but had merely kept hitting the mine itself—until they realized that if either of them had scored a bull's-eye, there would have been no one to tell the tale.

On the rest of our walk, too, Bohr and I amused ourselves by flinging stones at distant objects. On one occasion this activity again gave rise to a conversation about the powers of the imagination. I happened to see a telegraph pole quite a long distance away, almost too far to be reached with a stone. When the improbable nevertheless happened, and I hit it on my first attempt, Bohr became reflective: "If you had thought first about your aim, or about the correct angle of your arm and wrist, you wouldn't have had the least chance of scoring a hit. But since you were unreasonable enough to imagine that you could hit the target without special effort, why, you did it." We then had a lengthy discussion about the role of images and concepts in atomic physics. But more of this in another chapter.

We spent the night in a lonely inn at the edge of a forest in the northwestern part of the island. Next morning Bohr showed me around his country house in Tisvilde, in which we were later to have so many conversations about atomic physics. At this time of the year, the house was not yet ready to receive guests. Then we made our way back to Copenhagen and stopped briefly in Hilleröd, catching a glimpse of Frederiksborg Castle, a splendid Renaissance building in the Dutch style. It was surrounded by a lake and parklands and had obviously once served as a royal hunting lodge. I could sense that Bohr had a much greater liking for Hamlet's castle in Kronborg than for this rather trivial monument to the courtly life. No wonder, therefore, that the conversation turned back to atomic physics, a subject that was to fill so many of our future thoughts and perhaps the most important part of our lives.

5

Quantum Mechanics and
a Talk with Einstein (1925–1926)

During these critical years, atomic physics developed much as Niels Bohr had predicted it would during our walk over the Hain Mountain. The difficulties and inner contradictions that stood in the way of a true understanding of atoms and their stability seemed unlikely to be removed or even reduced—on the contrary, they became still more acute. All attempts to surmount them with the conceptual tools of the older physics appeared doomed to failure.

There was, for instance, the discovery by the American physicist, Arthur Holly Compton, that light (or more precisely X-rays) changes its wavelength when radiation is scattered by free electrons. This result could be explained by Einstein's hypothesis that light consists of small corpuscles or packets of energy, moving through space with great velocity and occasionally—e.g., during the process of scattering—colliding with an electron. On the other hand, there was a great deal of experimental evidence to suggest that the only basic difference between light and radio waves was that the former are of shorter length; in other words, that a light ray is a wave and not a stream of particles. Moreover, attempts by the Dutch physicist, Ornstein, to determine the intensity ratio of spectral lines in a so-called multiplet had produced very strange results. These ratios can be determined with the help of Bohr's theory. Now it appeared that, although the formulae derived from Bohr's theory were incorrect, a minor modification produced new formulae that fitted the experimental results. And so physicists gradually learned to adapt themselves to a host of difficulties. They became used to the fact that the concepts and models of classical physics were not rigorously applicable to processes on the atomic scale. On the other hand, they had come to appreciate that, by skillful use of the resulting freedom, they could, on occasion, guess the correct mathematical formulation of some of the details.

In the seminars run by Max Born in Göttingen during the summer of 1924, we had begun to speak of a new quantum mechanics that would one day oust the old Newtonian mechanics, and whose vague outlines could already be discerned here and there. Even during the subsequent winter term, which I once again spent in Copenhagen, trying to develop Kramers' theory of dispersion phenomena, our efforts were devoted not so much to deriving the correct mathematical relationships as to guessing them from similarities with the formulae of classical theory.

If I think back on the state of atomic theory in those months, I always remember a mountain walk with some friends from the Youth Movement, probably in the late autumn of 1924. It took us from Kreuth to Lake Achen. In the valley the weather was poor, and the mountains were veiled in clouds. During the climb, the mist had begun to close in upon us, and, after a time, we found ourselves in a confused jumble of rocks and undergrowth with no signs of a track. We decided to keep climbing, though we felt rather anxious about getting down again if anything went wrong. All at once the mist became so dense that we lost sight of one another completely, and could keep in touch only by shouting. At the same time it grew brighter overhead, and the light suddenly changed color. We were obviously under a patch of moving fog. Then, quite suddenly, we could see the edge of a steep rock face, straight ahead of us, bathed in bright sunlight. The next moment the fog had closed up again, but we had seen enough to take our bearings from the map. After a further ten minutes of hard climbing we were standing in the sun—at saddle height above the sea of fog. To the south we could see the peaks of the Sonnwend Mountains and beyond them the snowy tops of the Central Alps, and we all breathed a sigh of relief.

In atomic physics, likewise, the winter of 1924–1925 had obviously brought us to a realm where the fog was thick but where some light had begun to filter through and held out the promise of exciting new vistas.

In the summer term of 1925, when I resumed my research work at the University of Göttingen—since July 1924 I had been *Privatdozent* at that university—I made a first attempt to guess what formulae would enable one to express the line intensities of the hydrogen spectrum, using more or less the same methods that had proved so fruitful in my work with Kramers in Copenhagen. This attempt led to a dead end—I found myself in an impenetrable morass of complicated mathematical equations, with no way out. But the work helped to convince me of one thing: that one ought to ignore the problem of electron orbits inside the atom, and treat the frequencies and amplitudes associated with the line intensities as perfectly good substitutes. In any case, these magnitudes could be observed directly, and as my friend Otto had pointed out when expounding on Einstein's theory during our bicycle tour round Lake Walchensee, physicists must consider none but observable magnitudes when trying to solve the atomic puzzle. My attempt to apply this scheme to the hydrogen atom had come to grief on the complications of this particular problem. Accordingly, I looked for a simpler mathematical system and found it in the pendulum, whose oscillations could serve as a model for the molecular vibrations treated by atomic physics. My work along these lines was advanced rather than retarded by an unfortunate personal setback.

Toward the end of May 1925, I fell so ill with hay fever that I had to ask Born for fourteen days' leave of absence. I made straight for Heligoland, where I hoped to recover quickly in the bracing sea air, far from blossoms and meadows. On my arrival I must have looked quite a sight with my swollen face; in any case, my landlady took one look at me, concluded that I had been in a fight and promised to nurse me through the aftereffects. My room was on the second floor, and since the house was built high up on the southern edge of the rocky island, I had a glorious view over the village, and the dunes and the sea beyond. As I sat on my balcony, I had ample opportunity to reflect on Bohr's remark that part of infinity seems to lie within the grasp of those who look across the sea.

Apart from daily walks and long swims, there was nothing in Heligoland to distract me from my problem, and so I made much swifter progress than I would have done in Göttingen. A few days were enough to jettison all the mathematical ballast that invariably encumbers the beginning of such attempts, and to arrive at a simple formulation of my problem. Within a few days more, it had become clear to me what precisely had to take the place of the Bohr-Sommerfeld quantum conditions in an atomic physics working with none but observable magnitudes. It also became obvious that with this additional assumption I had introduced a crucial restriction into the theory. Then I noticed that there was no guarantee that the new mathematical scheme could be put into operation without contradictions. In particular, it was completely uncertain whether the principle of the conservation of energy would still apply, and I knew only too well that my scheme stood or fell by that principle.

Other than that, however, several calculations showed that the scheme seemed quite self-consistent. Hence I concentrated on demonstrating that the conservation law held, and one evening I reached the point where I was ready to determine the individual terms in the energy table, or, as we put it today, in the energy matrix, by what would now be considered an extremely clumsy series of calculations. When the first terms seemed to accord with the energy principle, I became rather excited, and I began to make countless arithmetical errors. As a result, it was almost three o'clock in the morning before the final result of my computations lay before me. The energy principle had held for all the terms, and I could no longer doubt the mathematical consistency and coherence of the kind of quantum mechanics to which my calculations pointed. At first, I was deeply alarmed. I had the feeling that, through the surface of atomic phenomena, I was looking at a strangely beautiful interior, and felt almost giddy at the thought that I now had to probe this wealth of mathematical structures nature had so generously spread out before me. I was far too excited to sleep, and so, as a new day dawned, I made for the southern tip of the island, where I had been longing to climb a rock jutting out into the sea. I now did so without too much trouble, and waited for the sun to rise.

What I saw during that night in Heligoland was admittedly not very much more than the sunlit rock edge I had glimpsed in the autumn of 1924, but when I reported my results to Wolfgang Pauli, generally my severest critic, he warmly encouraged me to continue along the path I had taken. In Göttingen, Max Born and Pascual Jordan took stock of the new possibilities, and in Cambridge the young English mathematician Paul Dirac developed his own methods for solving the problems involved, and after only a few months the concentrated efforts of these men led to the emergence of a coherent mathematical framework, one that promised to embrace all the multifarious aspects of atomic physics. Of the extremely intensive work which kept us breathless for a few months I shall say nothing here; instead, I shall report my talk with Albert Einstein following a lecture on the new quantum mechanics in Berlin.

At the time, the University of Berlin was considered the stronghold of physics in Germany, with such renowned figures as Planck, Einstein, von Laue and Nernst. It was here that Planck had discovered quantum theory and that Rubens had confirmed it by special measurements of thermal radiation; it was here that Einstein had formulated his general theory of relativity and his theory of gravitation in 1916. At the center of scientific life was the so-called physics colloquium, which probably went back to the time of Helmholtz and which was generally attended by the entire staff of the physics department. In the spring of 1926, I was invited to address this distinguished body on the new quantum mechanics, and since this was my first chance to meet so many famous men, I took good care to give a clear account of the concepts and mathematical foundations of what was then a most unconventional theory. I apparently managed to arouse Einstein's interest, for he invited me to walk home with him so that we might discuss the new ideas at greater length.

On the way, he asked about my studies and previous research. As soon as we were indoors, he opened the conversation with a question that bore on the philosophical background of my recent work. "What you have told us sounds extremely strange. You assume the existence of electrons inside the atom, and you are probably quite right to do so. But you refuse to consider their orbits, even though we can observe electron tracks in a cloud chamber. I should very much like to hear more about your reasons for making such strange assumptions."

"We cannot observe electron orbits inside the atom," I must have replied, "but the radiation which an atom emits during discharges enables us to deduce the frequencies and corresponding amplitudes of its electrons. After all, even in the older physics wave numbers and amplitudes could be considered substitutes for electron orbits. Now, since a good theory must be based on directly observable magnitudes, I thought it more fitting to restrict myself to these, treating them, as it were, as representatives of the electron orbits."

"But you don't seriously believe," Einstein protested, "that none but observable magnitudes must go into a physical theory?"

"Isn't that precisely what you have done with relativity?" I asked in some surprise. "After all, you did stress the fact that it is impermissible to speak of absolute time, simply because absolute time cannot be observed; that only clock readings, be it in the moving reference system or the system at rest, are relevant to the determination of time."

"Possibly I did use this kind of reasoning," Einstein admitted, "but it is nonsense all the same. Perhaps I could put it more diplomatically by saying that it may be heuristically useful to keep in mind what one has actually observed. But on principle, it is quite wrong to try founding a theory on observable magnitudes alone. In reality the very opposite happens. It is the theory which decides what we can observe. You must appreciate that observation is a very complicated process. The phenomenon under observation produces certain events in our measuring apparatus. As a result, further processes take place in the apparatus, which eventually and by complicated paths produce sense impressions and help us to fix the effects in our consciousness. Along this whole path—from the phenomenon to its fixation in our consciousness—we must be able to tell how nature functions, must know the natural laws at least in practical terms, before we can claim to have observed anything at all. Only theory, that is, knowledge of natural laws, enables us to deduce the underlying phenomena from our sense impressions. When we claim that we can observe something new, we ought really to be saying that, although we are about to formulate new natural laws that do not agree with the old ones, we nevertheless assume that the existing laws—covering the whole path from the phenomenon to our consciousness—function in such a way that we can rely upon them and hence speak of 'observations.'

"In the theory of relativity, for instance, we presuppose that, even in the moving reference system, the light rays traveling from the clock to the observer's eye behave more or less as we have always expected them to behave. And in your theory, you quite obviously assume that the whole mechanism of light transmission from the vibrating atom to the spectroscope or to the eye works just as one has always supposed it does, that is, essentially according to Maxwell's laws. If that were no longer the case, you could not possibly observe any of the magnitudes you call observable. Your claim that you are introducing none but observable magnitudes is therefore an assumption about a property of the theory that you are trying to formulate. You are, in fact, assuming that your theory does not clash with the old description of radiation phenomena in the essential points. You may well be right, of course, but you cannot be certain."

I was completely taken aback by Einstein's attitude, though I found his arguments convincing. Hence I said: "The idea that a good theory is no more than a condensation of observations in accordance with the principle of thought economy surely goes back to Mach, and it has, in fact, been said that your relativity theory makes decisive use of Machian concepts. But what you have just told me seems to indicate the very opposite. What am I to make of all this, or rather what do you yourself think about it?"

"It's a very long story, but we can go into it if you like. Mach's concept of thought economy probably contains part of the truth, but strikes me as being just a bit too trivial. Let me first of all produce a few arguments in its favor. We obviously grasp the world by way of our senses. Even when small children learn to speak and to think, they do so by recognizir₋ the possibility of describing highly complicated but somehow related sense impressions with a single word, for instance, the word 'ball.' They learn it from adults and get the satisfaction that they can make themselves understood. In other words, we may argue that the formation of the word, and hence of the concept, 'ball' is a kind of thought economy enabling the child to combine very complicated sense impressions in a simple way. Here Mach does not even enter into the question which mental or physical predispositions must be satisfied in man—or the small child—before the process of communication can be initiated. With animals, this process works considerably less effectively, as everyone knows, but we shan't talk about that now. Now Mach also thinks that the formation of scientific theories, however complex, takes place in a similar way. We try to order the phenomena, to reduce them to a simple form, until we can describe what may be a large number of them with the aid of a few simple concepts.

"All this sounds very reasonable, but we must nevertheless ask ourselves in what sense the principle of mental economy is being applied here. Are we thinking of psychological or of logical economy, or, again, are we dealing with the subjective or the objective side of the phenomena? When the child forms the concept 'ball,' does he introduce a purely psychological simplification in that he combines complicated sense impressions by means of this concept, or does this ball really exist? Mach would probably answer that the two statements express one and the same fact. But he would be quite wrong to do so. To begin with, the assertion 'The ball really exists' also contains a number of statements about possible sense impressions that may occur in the future. Now future possibilities and expectations make up a very important part of our reality, and must not be simply forgotten. Moreover, we ought to remember that inferring concepts and things from sense impressions is one of the basic presuppositions of all our thought. Hence, if we wanted to speak of nothing but sense impressions, we should have to rid ourselves of our language and thought. In other words, Mach rather neglects the fact that the world really exists, that our sense impressions are based on something objective.

"I have no wish to appear as an advocate of a naïve form of realism; I know that these are very difficult questions, but then I consider Mach's concept of observation also much too naïve. He pretends that we know perfectly well what the word 'observe' means, and thinks this exempts him from having to discriminate between 'objective' and 'subjective' phenomena. No wonder his principle has so suspiciously commercial a name: 'thought economy.' His idea of simplicity is much too subjective for me. In reality, the simplicity of natural laws is an objective fact as well, and the correct conceptual scheme must balance the subjective side of this simplicity with the objective. But that is a very difficult task. Let us rather return to your lecture.

"I have a strong suspicion that, precisely because of the problems we have just been discussing, your theory will one day get you into hot water. I should like to explain this in greater detail. When it comes to observation, you behave as if everything can be left as it was, that is, as if you could use the old descriptive language. In that case, however, you will also have to say: in a cloud chamber we can observe the path of the electrons. At the same time, you claim that there are no electron paths inside the atom. This is obvious nonsense, for you cannot possibly get rid of the path simply by restricting the space in which the electron moves."

I tried to come to the defense of the new quantum mechanics. "For the time being, we have no idea in what language we must speak about processes inside the atom. True, we have a mathematical language, that is, a mathematical scheme for determining the stationary states of the atom or the transition probabilities from one state to another, but we do not know—at least not in general—how this language is related to that of classical physics. And, of course, we need this connection if we are to apply this theory to experiments in the first place. For when it comes to experiments, we invariably speak in the traditional language. Hence I cannot really claim that we have 'understood' quantum mechanics. I assume that the mathematical scheme works, but no link with the traditional language has been established so far. And until that has been done, we cannot hope to speak of the path of the electron in the cloud chamber without inner contradictions. Hence it is probably much too early to solve the difficulties you have mentioned."

"Very well, I will accept that," Einstein said. "We shall talk about it again in a few years' time. But perhaps I may put another question to you. Quantum theory as you have expounded it in your lecture has two distinct faces. On the one hand, as Bohr himself has rightly stressed, it explains the stability of the atom; it causes the same forms to reappear time and again. On the other hand, it explains that strange discontinuity or inconstancy of nature which we observe quite clearly when we watch flashes of light on a scintillation screen. These two aspects are obviously connected. In your quantum mechanics you will have to take both into account, for instance when you speak of the emission of light by atoms. You can calculate the discrete energy values of the stationary states. Your theory can thus account for the stability of certain forms that cannot merge continuously into one another, but must differ by finite amounts and seem capable of permanent re-formation. But what happens during the emission of light? As you know, I suggested that, when an atom drops suddenly from one stationary energy value to the next, it emits the energy difference as an energy packet, a so-called light quantum. In that case, we have a particularly clear example of discontinuity. Do you think that my conception is correct? Or can you describe the transition from one stationary state to another in a more precise way?"

In my reply, I must have said something like this: "Bohr has taught me that one cannot describe this process by means of the traditional concepts, i.e., as a process in time and space. With that, of course, we have said very little, no more, in fact, than that we do not know. Whether or not I should believe in light quanta, I cannot say at this stage. Radiation quite obviously involves the discontinuous elements to which you refer as light quanta. On the other hand, there is a continuous element, which appears, for instance, in interference phenomena, and which is much more simply described by the wave theory of light. But you are of course quite right to ask whether quantum mechanics has anything new to say on these terribly difficult problems. I believe that we may at least hope that it will one day.

"I could, for instance, imagine that we should obtain an interesting answer if we considered the energy fluctuations of an atom during reactions with other atoms or with the radiation field. If the energy should change discontinuously, as we expect from your theory of light quanta, then the fluctuation, or, in more precise mathematical terms, the mean square fluctuation, would be greater than if the energy changed continuously. I am inclined to believe that quantum mechanics would lead to the greater value, and so establish the discontinuity. On the other hand, the continuous element, which appears in interference experiments, must also be taken into account. Perhaps one must imagine the transitions from one stationary state to the next as so many fade-outs in a film. The change is not sudden—one picture gradually fades while the next comes into focus so that, for a time, both pictures become confused and one does not know which is which. Similarly, there may well be an intermediate state in which we cannot tell whether an atom is in the upper or the lower state."

"You are moving on very thin ice," Einstein warned me. "For you are suddenly speaking of what we know about nature and no longer about what nature really does. In science we ought to be concerned solely with what nature does. It might very well be that you and I know quite different things about nature. But who would be interested in that? Perhaps you and I alone. To everyone else it is a matter of complete indifference. In other words, if your theory is right, you will have to tell me sooner or later what the atom does when it passes from one stationary state to the next."

"Perhaps," I may have answered. "But it seems to me that you are using language a little too strictly. Still, I do admit that everything that I might now say may sound like a cheap excuse. So let's wait and see how atomic theory develops."

Einstein gave me a skeptical look. "How can you really have so much faith in your theory when so many crucial problems remain completely unsolved?"

I must certainly have thought for a long time before I produced my answer. "I believe, just like you, that the simplicity of natural laws has an objective character, that it is not just the result of thought economy. If nature leads us to mathematical forms of great simplicity and beauty—by forms I am referring to coherent systems of hypotheses, axioms, etc.—to forms that no one has previously encountered, we cannot help thinking that they are 'true,' that they reveal a genuine feature of nature. It may be that these forms also cover our subjective relationship to nature, that they reflect elements of our own thought economy. But the mere fact that we could never have arrived at these forms by ourselves, that they were revealed to us by nature, suggests strongly that they must be part of reality itself, not just of our thoughts about reality.

"You may object that by speaking of simplicity and beauty I

am introducing aesthetic criteria of truth, and I frankly admit that I am strongly attracted by the simplicity and beauty of the mathematical schemes with which nature presents us. You must have felt this, too: the almost frightening simplicity and wholeness of the relationships which nature suddenly spreads out before us and for which none of us was in the least prepared. And this feeling is something completely different from the joy we feel when we have done a set task particularly well. That is one reason why I hope that the problems we have been discussing will be solved in one way or another. In the present case, the simplicity of the mathematical scheme has the further consequence that it ought to be possible to think up many experiments whose results can be predicted from the theory. And if the actual experiments should bear out the predictions, there is little doubt but that the theory reflects nature accurately in this particular realm."

"Control by experiment," Einstein agreed, "is, of course, an essential prerequisite of the validity of any theory. But one can't possibly test everything. That is why I am so interested in your remarks about simplicity. Still, I should never claim that I really understood what is meant by the simplicity of natural laws."

After talking about the role of truth criteria in physics for quite a bit longer, I took my leave. I next met Einstein a year and a half later, at the Solvay Congress in Brussels, where the epistemological and philosophical bases of quantum theory once again formed the subject of the most exciting discussions.

6

Fresh Fields (1926–1927)

If I were asked what was Christopher Columbus' greatest achievement in discovering America, my answer would not be that he took advantage of the spherical shape of the earth to get to India by the western route—this idea had occurred to others before him—or that he prepared his expedition meticulously and rigged his ships most expertly—that, too, others could have done equally well. His most remarkable feat was the decision to leave the known regions of the world and to sail westward, far beyond the point from which his provisions could have got him back home again.

In science, too, it is impossible to open up new territory unless one is prepared to leave the safe anchorage of established doctrine and run the risk of a hazardous leap forward. With his relativity theory, Einstein had abandoned the concept of simultaneity, which was part of the solid ground of traditional physics, and, in so doing, outraged many leading physicists and philosophers and turned them into bitter opponents. In general, scientific progress calls for no more than the absorption and elaboration of new ideas—and this is a call most scientists are happy to heed. However, when it comes to entering new territory, the very structure of scientific thought may have to be changed, and that is far more than most men are prepared to do. How great their reluctance could be had been brought home to me at the Leipzig Congress, and I fully expected that similar obstacles would be placed in the path of atomic physics.

During the first few months of 1926, at about the same time that I delivered my lecture in Berlin, Göttingen first became

familiar with the work of the Viennese physicist, Erwin Schrödinger, who was approaching atomic theory from an entirely
fresh side. The year before, Louis de Broglie in France had
drawn attention to the fact that the strange wave-particle dualism which, at the time, seemed to prevent a rational explanation
of light phenomena might be equally involved in the behavior of
matter, for instance of electrons. Schrödinger developed this idea
further and, by means of a new wave equation, formulated the
law governing the propagation of material waves under the influence of an electromagnetic field. In Schrödinger's model, the
stationary states of an atomic shell are compared with the stationary vibrations of a system, for instance of a vibrating string,
except that all the magnitudes normally considered as energies
of the stationary states are treated as frequencies of the stationary vibrations. The results Schrödinger obtained in this
way fitted in very well with the new quantum mechanics, and
Schrödinger quickly succeeded in proving that his own wave
mechanics was mathematically equivalent to quantum mechanics; in other words, that the two were but different mathematical
formulations of the same structures. Needless to say, we were
delighted by this new development, for it greatly strengthened
our confidence in the correctness of the new mathematical formulation. Moreover, Schrödinger's procedure lent itself readily to
the simplification of calculations that had severely strained the
powers of quantum mechanics.

Unfortunately, however, the physical interpretation of the
mathematical scheme presented us with grave problems. Schrödinger believed that, by associating particles with material waves,
he had found a way of clearing the obstacles that had so long
blocked the path of quantum theory. According to him, these
material waves were fully comparable to such processes in space
and time as electromagnetic or sound waves. Such obscure ideas
as quantum jumps would completely disappear. I had no faith in
a theory that ran completely counter to our Copenhagen conception and was disturbed to see that so many physicists greeted
precisely this part of Schrödinger's doctrine with a sense of
liberation. The many talks I had had with Niels Bohr, Wolfgang
Pauli and many others over the years had convinced me that it
was impossible to build up a descriptive time-space model of

interatomic processes—the discontinuous element Einstein had mentioned to me in Berlin as a characteristic feature of atomic phenomena saw to that. Admittedly, this was no more than a negative feature, and we were still a long way from a complete physical interpretation of quantum mechanics, yet we were certain that we must get away from the idea of objective processes in time and space.

Now Schrödinger's interpretation—and this was its great novelty —simply denied the existence of these discontinuities. Thus when an atom passes from one stationary state to the next, it was no longer said to change its energy suddenly and to radiate the difference in the form of an Einsteinian light quanta. Radiation was the result of quite a different process, namely, of the simultaneous excitation of two stationary material vibrations whose interference gives rise to the emission of electromagnetic waves, e.g., light. This hypothesis seemed to me too good to be true, and I mustered what arguments I could to show that discontinuities · were a fact of life, however inconvenient. The simplest argument was, of course, Planck's radiation formula, whose empirical correctness no one could doubt and which, after all, had led Planck to his discrete energy quanta.

Toward the end of the 1926 summer term, Sommerfeld invited Schrödinger to address the Munich seminar. I had been working in Copenhagen once again and had familiarized myself with Schrödinger's methods by applying them to the study of the helium atom. I had finished the work while taking a brief holiday on Lake Mjösa in Norway, had stuffed the manuscript into my rucksack and had set out on unmade paths from Gudbrandsdal, across several mountain chains, to Sogne Fjord. After a short stay in Copenhagen, I finally went on to Munich, where I intended to spend the rest of the vacation with my parents—and so I could be present at Schrödinger's lecture, and discuss his theory with him in person. The audience included the director of the Institute for Experimental Physics in the University of Munich, Wilhelm Wien, who was extremely skeptical of Sommerfeld's "atomysticism."

Schrödinger first of all explained the mathematical principles of wave mechanics by using the hydrogen atom as an illustration. All of us were delighted to see his elegant and simple solution by conventional methods of a problem that Wolfgang Pauli had been able to solve only with great difficulty using quantum mechanics. Unfortunately, Schrödinger went on to discuss his own intepretation of wave mechanics, and his arguments left me quite unconvinced. During the subsequent discussion, I therefore raised a number of objections, and, in particular, pointed out that Schrödinger's conception would not even help explain Planck's radiation law. For this I was taken to task by Wilhelm Wien, who told me rather sharply that while he understood my regrets that quantum mechanics was finished, and with it all such nonsense as quantum jumps, etc., the difficulties I had mentioned would undoubtedly be solved by Schrödinger in the very near future. Schrödinger himself was not quite so certain in his own reply, but he, too, remained convinced that it was only a question of time before my objections would be removed. My arguments had clearly failed to impress anyone—even Sommerfeld, who felt most kindly toward me, succumbed to the persuasive force of Schrödinger's mathematics.

And so I went home rather sadly. It must have been that same evening that I wrote to Niels Bohr about the unhappy outcome of the discussion. Perhaps it was as a result of this letter that he invited Schrödinger to spend part of September in Copenhagen. Schrödinger agreed, and I, too, sped back to Denmark.

Bohr's discussions with Schrödinger began at the railway station and were continued daily from early morning until late at night. Schrödinger stayed in Bohr's house so that nothing would interrupt the conversations. And although Bohr was normally most considerate and friendly in his dealings with people, he now struck me as an almost remorseless fanatic, one who was not prepared to make the least concession or grant that he could ever be mistaken. It is hardly possible to convey just how passionate the discussions were, just how deeply rooted the convictions of each, a fact that marked their every utterance. All I can hope to do here is to produce a very pale copy of conversations in which two men were fighting for their particular interpretation of the new mathematical scheme with all the powers at their command.

Schrödinger: "Surely you realize that the whole idea of quantum jumps is bound to end in nonsense. You claim first of all that if an atom is in a stationary state, the electron revolves periodically but does not emit light, when, according to Maxwell's theory, it must. Next, the electron is said to jump from one orbit to the next and to emit radiation. Is this jump supposed to be gradual or sudden? If it is gradual, the orbital frequency and energy of the electron must change gradually as well. But in that case, how do you explain the persistence of fine spectral lines? On the other hand, if the jump is sudden, Einstein's idea of light quanta will admittedly lead us to the right wave number, but then we must ask ourselves how precisely the electron behaves during the jump. Why does it not emit a continuous spectrum, as electromagnetic theory demands? And what laws govern its motion during the jump? In other words, the whole idea of quantum jumps is sheer fantasy."

Bohr: "What you say is absolutely correct. But it does not prove that there are no quantum jumps. It only proves that we cannot imagine them, that the representational concepts with which we describe events in daily life and experiments in classical physics are inadequate when it comes to describing quantum jumps. Nor should we be surprised to find it so, seeing that the processes involved are not the objects of direct experience."

Schrödinger: "I don't wish to enter into long arguments about the formation of concepts; I prefer to leave that to the philosophers. I wish only to know what happens inside an atom. I don't really mind what language you choose to discuss it. If there are electrons in the atom, and if these are particles—as all of us believe—then they must surely move in some way. Right now I am not concerned with a precise description of this motion, but it ought to be possible to determine in principle how they behave in the stationary state or during the transition from one state to the next. But from the mathematical form of wave or quantum mechanics alone it is clear that we cannot expect reasonable answers to these questions. The moment, however, that we change the picture and say that there are no discrete electrons, only electron waves or waves of matter, then everything looks quite different. We no longer wonder about the fine lines. The emission of light is as easily explained as the transmission of radio waves through the aerial of the transmitter, and what seemed to be insoluble contradictions have suddenly disappeared."

Bohr: "I beg to disagree. The contradictions do not disappear; they are simply pushed to one side. You speak of the emission of light by the atom or more generally of the interaction between the atom and the surrounding radiation field, and you think that all the problems are solved once we assume that there are material waves but no quantum jumps. But just take the case of thermodynamic equilibrium between the atom and the radiation field—remember, for instance, the Einsteinian derivation of Planck's radiation law. This derivation demands that the energy of the atom should assume discrete values and change discontinuously from time to time; discrete values for the frequencies cannot help us here. You can't seriously be trying to cast doubt on the whole basis of quantum theory!"

Schrödinger: "I don't for a moment claim that all these relationships have been fully explained. But then you, too, have so far failed to discover a satisfactory physical interpretation of quantum mechanics. There is no reason why the application of thermodynamics to the theory of material waves should not yield a satisfactory explanation of Planck's formula as well—an explanation that will admittedly look somewhat different from all previous ones."

Bohr: "No, there is no hope of that at all. We have known what Planck's formula means for the past twenty-five years. And, quite apart from that, we can see the inconstancies, the sudden jumps in atomic phenomena quite directly, for instance when we watch sudden flashes of light on a scintillation screen or the sudden rush of an electron through a cloud chamber. You cannot simply ignore these observations and behave as if they did not exist at all."

Schrödinger: "If all this damned quantum jumping were really here to stay, I should be sorry I ever got involved with quantum theory."

Bohr: "But the rest of us are extremely grateful that you did; your wave mechanics has contributed so much to mathematical clarity and simplicity that it represents a gigantic advance over all previous forms of quantum mechanics."

And so the discussions continued day and night. After a few days Schrödinger fell ill, perhaps as a result of his enormous effort; in any case, he was forced to keep to his bed with a feverish cold. While Mrs. Bohr nursed him and brought in tea and cake, Niels Bohr kept sitting on the edge of the bed talking at Schrödinger: "But you must surely admit that . . ." No real understanding could be expected since, at the time, neither side was able to offer a complete and coherent interpretation of quantum mechanics. For all that, we in Copenhagen felt convinced toward the end of Schrödinger's visit that we were on the right track, though we fully realized how difficult it would be to convince even leading physicists that they must abandon all attempts to construct perceptual models of atomic processes.

During the next few months the physical interpretation of quantum mechanics was the central theme of all conversations between Bohr and myself. I was then living on the top floor of the Institute, in a cozy little attic flat with slanting walls and windows overlooking the trees at the entrance to Faelled Park. Bohr would often come into my attic late at night, and we constructed all sorts of imaginary experiments to see whether we had really grasped the theory. In so doing, we discovered that the two of us were trying to resolve the difficulties in rather different ways. Bohr was trying to allow for the simultaneous existence of both particle and wave concepts, holding that, though the two were mutually exclusive, both together were needed for a complete description of atomic processes. I disliked this approach. I wanted to start from the fact that quantum mechanics as we then knew it already imposed a unique physical interpretation of some magnitudes occurring in it—for instance, the time averages of energy, momentum, fluctuations, etc.—so that it looked very much as if we no longer had any freedom with respect to that interpretation. Instead, we would have to try to derive the correct general interpretation by strict logic from the ready-to-hand, more special interpretation.

For that reason I was—certainly quite wrongly—rather unhappy about a brilliant piece of work Max Born had done in Göttingen. In it, he had treated collisions by Schrödinger's method and assumed that the square of the Schrödinger wave function measures, in each point of space and at every instant, the probability of finding an electron in this point at that instant. I fully agreed with Born's thesis as such, but disliked the fact that it looked as if we still had some freedom of interpretation; I was firmly convinced that Born's thesis itself was the necessary consequence of the fixed interpretation of special magnitudes in quantum mechanics. This conviction was strengthened further by two highly informative mathematical studies by Dirac and Jordan.

Luckily, at the end of our talks, Bohr and I would generally come to the same conclusions about particular physical experiments, so that there was good reason to think that our divergent efforts might yet lead to the same result. On the other hand, neither of us could tell how so simple a phenomenon as the trajectory of an electron in a cloud chamber could be reconciled with the mathematical formulations of quantum or wave mechanics. Such concepts as trajectories or orbits did not figure in quantum mechanics, and wave mechanics could only be reconciled with the existence of a densely packed beam of matter if the beam spread over areas much larger than the diameter of an electron.

Since our talks often continued till long after midnight, and did not produce a satisfactory conclusion despite protracted efforts over several months, both of us became utterly exhausted and rather tense. Hence Bohr decided in February 1927 to go skiing in Norway, and I was quite glad to be left behind in Copenhagen, where I could think about these hopelessly complicated problems undisturbed. I now concentrated all my efforts on the mathematical representation of the electron path in the cloud chamber, and when I realized fairly soon that the obstacles before me were quite insurmountable, I began to wonder whether we might not have been asking the wrong sort of question all along. But where had we gone wrong? The path of the electron through the cloud chamber obviously existed; one could easily observe it. The mathematical framework of quantum mechanics existed as well, and was much too convincing to allow for any changes. Hence it ought to be possible to establish a connection between the two, hard though it appeared to be.

It must have been one evening after midnight when I suddenly remembered my conversation with Einstein and particularly his statement, "It is the theory which decides what we can observe." I was immediately convinced that the key to the gate that had been closed for so long must be sought right here. I decided to go on a nocturnal walk through Faelled Park and to think further about the matter. We had always said so glibly that the path of the electron in the cloud chamber could be observed. But perhaps what we really observed was something much less. Perhaps we merely saw a series of discrete and ill-defined spots through which the electron had passed. In fact, all we do see in the cloud chamber are individual water droplets which must certainly be much larger than the electron. The right question should therefore be: Can quantum mechanics represent the fact that an electron finds itself approximately in a given place and that it moves approximately with a given velocity, and can we make these approximations so close that they do not cause experimental difficulties?

A brief calculation after my return to the Institute showed that one could indeed represent such situations mathematically, and that the approximations are governed by what would later be called the uncertainty principle of quantum mechanics: the product of the uncertainties in the measured values of the position and momentum (i.e., the product of mass and velocity) cannot be smaller than Planck's constant. This formulation, I felt, established the much-needed bridge between the cloud-chamber observations and the mathematics of quantum mechanics. True, it had still to be proved that any experiment whatsoever was bound to set up situations satisfying the uncertainty principle, but this struck me as plausible a priori, since the processes involved in the experiment or the observation had necessarily to satisfy the laws of quantum mechanics. On this presupposition, experiments are unlikely to produce situations that do not accord with quantum mechanics. "It is the theory which decides what we can observe." I resolved to prove this by calculations based on simple experiments during the next few days.

Here, too, I was helped by the memory of a conversation I once had with Burkhard Drude, a fellow student in Göttingen. When discussing the difficulties involved in the concept of electron orbits, he had said that it ought to be possible, in principle, to construct a microscope of extraordinarily high resolving power in which one could see or photograph the electron paths inside the atom. Such a microscope would not, of course, work with ordinary light rays, but perhaps with gamma rays. Now this ran counter to my hypothesis, according to which not even the best microscope could cross the limits set by the uncertainty principle. Hence I had to demonstrate that the principle was obeyed even in this case. This I managed to do, and the proof strengthened my confidence in the consistency of the new interpretation. After a few more calculations of this kind, I sat down and summarized my results in a long letter to Wolfgang Pauli. His encouraging reply from Hamburg cheered me considerably.

Then Niels Bohr returned from his skiing holiday, and we had a fresh round of difficult discussions. For Bohr, too, had pursued his own ideas on wave-corpuscle dualism. Central to his thought was the concept of complementarity, which he had just introduced to describe a situation in which it is possible to grasp one and the same event by two distinct modes of interpretation. These two modes are mutually exclusive, but they also complement each other, and it is only through their juxtaposition that the perceptual content of a phenomenon is fully brought out. At first, Bohr raised a number of objections against the uncertainty principle, which he probably considered too special a case of the general rule of complementarity. But he soon afterward realized —manfully assisted by the Swedish physicist, Oskar Klein, who was also working in Copenhagen—that there was no serious difference between the two interpretations, and that all that mattered now was to represent the facts in such a way that despite their novelty they could be grasped and accepted by all physicists.

The matter was thrashed out in the autumn of 1927 at two physics conferences: the General Physics Congress in Como, at which Bohr gave a comprehensive account of the new situation, and the Solvay Congress in Brussels. In accordance with the wishes of the Solvay Foundation, the latter was attended by a small group of specialists anxious to discuss the problems of quantum theory in detail. We all stayed at the same hotel, and the keenest arguments took place, not in the conference hall but during the hotel meals. Bohr and Einstein were in the thick of it all. Einstein was quite unwilling to accept the fundamentally statistical character of the new quantum theory. Needless to say, he had no objections against probability statements whenever a particular system was not known in every last detail—after all, the old statistical mechanics and thermodynamics had been based on just such statements. However, Einstein would not admit that it was impossible, even in principle, to discover all the partial facts needed for the complete description of a physical process. "God does not throw dice" was a phrase we often heard from his lips in these discussions. And so he refused point-blank to accept the uncertainty principle, and tried to think up cases in which the principle would not hold.

The discussion usually started at breakfast, with Einstein serving us up with yet another imaginery experiment by which he thought he had definitely refuted the uncertainty principle. We would at once examine his fresh offering, and on the way to the conference hall, to which I generally accompanied Bohr and Einstein, we would clarify some of the points and discuss their relevance. Then, in the course of the day, we would have further discussions on the matter, and, as a rule, by suppertime we would have reached the point where Niels Bohr could prove to Einstein that even his latest experiment failed to shake the uncertainty principle. Einstein would look a bit worried, but by next morning he was ready with a new imaginery experiment more complicated than the last, and this time, so he avowed, bound to invalidate the uncertainty principle. This attempt would fare no better by evening, and after the same game had been continued for a few days, Einstein's friend Paul Ehrenfest, a physicist from Leyden in Holland, said: "Einstein, I am ashamed of you; you are arguing against the new quantum theory just as your opponents argue about relativity theory." But even this friendly admonition went unheard.

Once again it was driven home to me how terribly difficult it is to give up an attitude on which one's entire scientific approach and career have been based. Einstein had devoted his life to probing into that objective world of physical processes which runs its course in space and time, independent of us, according to firm laws. The mathematical symbols of theoretical physics were also symbols of this objective world and as such enabled physicists to make statements about its future behavior. And now it was being asserted that, on the atomic scale, this objective world of time and space did not even exist and that the mathematical symbols of theoretical physics referred to possibilities rather than to facts. Einstein was not prepared to let us do what, to him, amounted to pulling the ground from under his feet. Later in life, also, when quantum theory had long since become an integral part of modern physics, Einstein was unable to change his attitude—at best, he was prepared to accept the existence of quantum theory as a temporary expedient. "God does not throw dice" was his unshakable principle, one that he would not allow anybody to challenge. To which Bohr could only counter with: "Nor is it our business to prescribe to God how He should run the world."

7

Science and Religion (1927)

One evening during the Solvay Conference, some of the younger members stayed behind in the lounge of our hotel. This group included Wolfgang Pauli and myself, and was soon afterward joined by Paul Dirac. One of us said: "Einstein keeps talking about God: what are we to make of that? It is extremely difficult to imagine that a scientist like Einstein should have such strong ties with a religious tradition."

"Not so much Einstein as Max Planck," someone objected. "From some of Planck's utterances it would seem that he sees no contradiction between religion and science, indeed that he believes the two are perfectly compatible."

I was asked what I knew of Planck's views on the subject, and what I thought myself. I had spoken to Planck on only a few occasions, mostly about physics and not about general questions, but I was acquainted with some of Planck's close friends, who had told me a great deal about his attitude.

"I assume," I must have replied, "that Planck considers religion and science compatible because, in his view, they refer to quite distinct facets of reality. Science deals with the objective, material world. It invites us to make accurate statements about objective reality and to grasp its interconnections. Religion, on the other hand, deals with the world of values. It considers what ought to be or what we ought to do, not what is. In science we are concerned to discover what is true or false; in religion with what is good or evil, noble or base. Science is the basis of technology, religion the basis of ethics. In short, the conflict between the two, which has been raging since the eighteenth century, seems founded on a misunderstanding, or, more precisely, on a confusion of the images and parables of religion with scientific statements. Needless to say, the result makes no sense at all. This view, which I know so well from my parents, associates the two realms with the objective and subjective aspects of the world respectively. Science is, so to speak, the manner in which we confront, in which we argue about, the objective side of reality. Religious faith, on the other hand, is the expression of the subjective decisions that help us choose the standards by which we propose to act and live. Admittedly, we generally make these decisions in accordance with the attitudes of the group to which we belong, be it our family, nation or culture. Our decisions are strongly influenced by educational and environmental factors, but in the final analysis they are subjective and hence not governed by the 'true or false' criterion. Max Planck, if I understand him rightly, has used this freedom and come down squarely on the side of the Christian tradition. His thoughts and actions, particularly as they affect his personal relationships, fit perfectly into the framework of this tradition, and no one will respect him the less for it. As far as he is concerned, therefore, the two realms—the objective and the subjective facets of the world—are quite separate, but I must confess that I myself do not feel altogether happy about this separation. I doubt whether human societies can live with so sharp a distinction between knowledge and faith."

Wolfgang shared my concern. "It's all bound to end in tears," he said. "At the dawn of religion, all the knowledge of a particular community fitted into a spiritual framework, based largely on religious values and ideas. The spiritual framework itself had to be within the grasp of the simplest member of the community, even if its parables and images conveyed no more than the vaguest hint as to their underlying values and ideas. But if he himself is to live by these values, the average man has to be convinced that the spiritual framework embraces the entire wisdom of his society. For 'believing' does not to him mean 'taking for granted,' but rather 'trusting in the guidance' of accepted values. That is why society is in such danger whenever fresh knowledge threatens to explode the old spiritual forms. The complete separation of knowledge and faith can at best be an emergency measure, afford some temporary relief. In Western culture, for instance, we may well reach the point in the not too distant future where the parables and images of the old religions will have lost their persuasive force even for the average person; when that happens, I am afraid that all the old ethics will collapse like a house of cards and that unimaginable horrors will be perpetrated. In brief, I cannot really endorse Planck's philosophy, even if it is logically valid and even though I respect the human attitudes to which it gives rise.

"Einstein's conception is closer to mine. His God is somehow involved in the immutable laws of nature. Einstein has a feeling for the central order of things. He can detect it in the simplicity of natural laws. We may take it that he felt this simplicity very strongly and directly during his discovery of the theory of relativity. Admittedly, this is a far cry from the contents of religion. I don't believe Einstein is tied to any religious tradition, and I rather think the idea of a personal God is entirely foreign to him. But as far as he is concerned there is no split between science and religion: the central order is part of the subjective as well as the objective realm, and this strikes me as being a far better starting point."

"A starting point for what?" I asked. "If you consider man's attitude to the central order a purely personal matter, then you may agree with Einstein's view, but then you must also concede that nothing at all follows from this view."

"Perhaps it does," Wolfgang replied. "The development of science during the past two centuries has certainly changed man's thinking, even outside the Christian West. Hence it matters quite a bit what physicists think. And it was precisely the idea of an objective world running its course in time and space according to strict causal laws that produced a sharp clash between science and the spiritual formulations of the various religions. If science goes beyond this strict view—and it has done just that with relativity theory and is likely to go even further with quantum theory—then the relationship between science and the contents religions try to express must change once again. Perhaps science, by revealing the existence of new relationships during the past thirty years, may have lent our thought much greater depth. The concept of complementarity, for instance, which

Niels Bohr considers so crucial in the interpretation of quantum theory, was by no means unknown to philosophers, even if they did not express it so succinctly. However, its very appearance in the exact sciences has constituted a decisive change: the idea of material objects that are completely independent of the manner in which we observe them proved to be nothing but an abstract extrapolation, something that has no counterpart in nature. In Asiatic philosophy and Eastern religions we find the complementary idea of a pure subject of knowledge, one that confronts no object. This idea, too, will prove an abstract extrapolation, corresponding to no spiritual or mental reality. If we think about the wider context, we may in the future be forced to keep a middle course between these extremes, perhaps the one charted by Bohr's complementarity concept. Any science that adapts itself to this form of thinking will not only be more tolerant of the different forms of religion, but, having a wider over-all view, may also contribute to the world of values."

Paul Dirac had joined us in the meantime. He had only just turned twenty-five, and had little time for tolerance. "I don't know why we are talking about religion," he objected. "If we are honest—and scientists have to be—we must admit that religion is a jumble of false assertions, with no basis in reality. The very idea of God is a product of the human imagination. It is quite understandable why primitive people, who were so much more exposed to the overpowering forces of nature than we are today, should have personified these forces in fear and trembling. But nowadays, when we understand so many natural processes, we have no need for such solutions. I can't for the life of me see how the postulate of an Almighty God helps us in any way. What I do see is that this assumption leads to such unproductive questions as why God allows so much misery and injustice, the exploitation of the poor by the rich and all the other horrors He might have prevented. If religion is still being taught, it is by no means because its ideas still convince us, but simply because some of us want to keep the lower classes quiet. Quiet people are much easier to govern than clamorous and dissatisfied ones. They are also very much easier to exploit. Religion is a kind of opium that allows a nation to lull itself into wishful dreams and so forget the injustices that are being perpetrated against the people. Hence the close alliance between those two great political forces, the State and the Church. Both need the illusion that a kindly God rewards—in heaven if not on earth—all those who have not risen up against injustice, who have done their duty quietly and uncomplainingly. That is precisely why the honest assertion that God is a mere product of the human imagination is branded as the worst of all mortal sins."

"You are simply judging religion by its political.abuses," I objected, "and since most things in this world can be abused— even the Communist ideology which you recently propounded— all such judgments are inadmissible. After all, there will always be human societies, and these must find a common language in which they can speak about life and death, and about the wider context in which their lives are set. The spiritual forms that have developeḑ historically out of this search for a common language must have had a great persuasive force—how else could so many people have lived by them for so many centuries? Religion can't be dismissed as simply as all that. But perhaps you are drawn to another religion, such as the old Chinese, in which the idea of a personal God does not occur?"

"I dislike religious myths on principle," Paul Dirac replied, "if only because the myths of the different religions contradict one another. After all, it was purely by chance that I was born in Europe and not in Asia, and that is surely no criterion for judging what is true or what I ought to believe. And I can only believe what is true. As for right action, I can deduce it by reason alone from the situation in which I find myself: I live in society with others, to whom, on principle, I must grant the same rights I claim for myself. I must simply try to strike a fair balance; no more can be asked of me. All this talk about God's will, about sin and repentance, about a world beyond by which we must direct our lives, only serves to disguise the sober truth. Belief in God merely encourages us to think that God wills us to submit to a higher force, and it is this idea which helps to preserve social structures that may have been perfectly good in their day but no longer fit the modern world. All your talk of a wider context and the like strikes me as quite unacceptable. Life, when all is said and done, is just like science: we come up against difficulties and have to solve them. And we can never solve more than one difficulty at a time; your wider context is nothing but a mental superstructure added a posteriori."

And so the discussion continued, and we were all of us surprised to notice that Wolfgang was keeping so silent. He would pull a long face or smile rather maliciously from time to time, but he said nothing. In the end, we had to ask him to tell us what he thought. He seemed a little surprised and then said: "Well, our friend Dirac, too, has a religion, and its guiding principle is: 'There is no God and Dirac is His prophet.'" We all laughed, including Dirac, and this brought our evening in the hotel lounge to a close.

Some time later, probably in Copenhagen, I told Niels about our conversation. He immediately jumped to the defense of the youngest member of our circle. "I consider it marvelous," he said, "that Paul should be so uncompromising in his defense of all that can be expressed in clear and logical language. He believes that what can be said at all can be said clearly—or, as Wittgenstein put it, that 'whereof one cannot speak thereof one must be silent.' Whenever Dirac sends me a manuscript, the writing is so neat and free of corrections that merely looking at it is an aesthetic pleasure. If I suggest even minor changes, Paul becomes terribly unhappy and generally changes nothing at all. His work is, in any case, quite brilliant. Recently the two of us went to an exhibition which included a glorious gray-blue seascape by Manet. In the foreground was a boat, and beside it, in the water, a dark gray spot, whose meaning was not quite clear. Dirac said, 'This spot is not admissible.' A strange way of looking at art, but he was probably quite right. In a good work of art, just as in a good piece of scientific work, every detail must be laid down quite unequivocally; there can be no room for mere accident.

"Still, religion is rather a different matter. I feel very much like Dirac: the idea of a personal God is foreign to me. But we ought to remember that religion uses language in quite a different way from science. The language of religion is more closely related to the language of poetry than to the language of science. True, we are inclined to think that science deals with information about objective facts, and poetry with subjective feelings. Hence we conclude that if religion does indeed deal with objective truths, it ought to adopt the same criteria of truth as science. But I myself find the division of the world into an objective and a subjective side much too arbitrary. The fact that religions through the ages have spoken in images, parables and paradoxes means simply that there are no other ways of grasping the reality to which they refer. But that does not mean that it is not a genuine reality. And splitting this reality into an objective and a subjective side won't get us very far.

"That is why I consider those developments in physics during the last decades which have shown how problematical such concepts as 'objective' and 'subjective' are, a great liberation of thought. The whole thing started with the theory of relativity. In the past, the statement that two events are simultaneous was considered an objective assertion, one that could be communicated quite simply and that was open to verification by any observer. Today we know that 'simultaneity' contains a subjective element, inasmuch as two events that appear simultaneous to an observer at rest are not necessarily simultaneous to an observer in motion. However, the relativistic description is also objective inasmuch as every observer can deduce by calculation what the other observer will perceive or has perceived. For all that, we have come a long way from the classical ideal of objective descriptions.

"In quantum mechanics the departure from this ideal has been even more radical. We can still use the objectifying language of classical physics to make statements about observable facts. For instance, we can say that a photographic plate has been blackened, or that cloud droplets have formed. But we can say nothing about the atoms themselves. And what predictions we base on such findings depend on the way we pose our experimental question, and here the observer has freedom of choice. Naturally, it still makes no difference whether the observer is a man, an animal or a piece of apparatus, but it is no longer possible to make predictions without reference to the observer or the means of observation. To that extent, every physical process may be said to have objective and subjective features. The objective world of nineteenth-century science was, as we know today, an ideal, limiting case, but not the whole reality. Admittedly, even in our future encounters with reality we shall have to distinguish between the objective and the subjective side, to make a division between the two. But the location of the separation may depend on the way things are looked at; to a certain extent it can be chosen at will. Hence I can quite understand why we cannot speak about the content of religion in an objectifying language. The fact that different religions try to express this content in quite distinct spiritual forms is no real objection. Perhaps we ought to look upon these different forms as complementary descriptions which, though they exclude one another, are needed to convey the rich possibilities flowing from man's relationship with the central order."

"If you distinguish so sharply between the languages of religion, science and art," I asked, "what meaning do you attach to such apodictic statements as 'There is a living God' or 'There is an immortal soul'? What is the meaning of 'there is' in this type of language? Science, like Dirac, objects to such formulations. Let me illustrate the epistemological side of the problem by means of the following analogy:

"Mathematicians, as everyone knows, work with an imaginary unit, the square root of -1, called i. We know that i does not figure among the natural numbers. Nevertheless, important branches of mathematics, for instance the theory of analytical functions, are based on this imaginary unit, that is, on the fact that $\sqrt{-1}$ exists after all. Would you agree that the statement 'There is a $\sqrt{-1}$' means nothing else than 'There are important mathematical relations that are most simply represented by the introduction of the $\sqrt{-1}$ concept'? And yet these relations would exist even without it. That is precisely why this type of mathematics is so useful even in science and technology. What is decisive, for instance, in the theory of functions, is the existence of important mathematical laws governing the behavior of pairs of continuous variables. These relations are rendered more comprehensible by the introduction of the abstract concept of $\sqrt{-1}$, although that concept is not basically needed for our understanding, and although it has no counterpart among the natural numbers. An equally abstract concept is that of infinity, which also plays a very important role in modern mathematics. It, too, has no correlate, and moreover raises grave problems. In short, mathematics introduces ever higher stages of abstraction that help us attain a coherent grasp of ever wider realms. To get back to our original question, is it correct to look upon the religious 'there is' as just another, though different, attempt to reach ever higher levels of abstraction? An attempt to facilitate our understanding of universal connections? After all, the connections themselves are real enough, no matter into what spiritual forms we try to fit them.''

"With respect to the epistemological side of the problem, your comparison may pass," Bohr replied. "But in other respects it is quite inadequate. In mathematics we can take our inner distance from the content of our statements. In the final analysis mathematics is a mental game that we can play or not play as we choose. Religion, on the other hand, deals with ourselves, with our life and death; its promises are meant to govern our actions and thus, at least indirectly, our very existence. We cannot just look at them impassively from the outside. Moreover, our attitude to religious questions cannot be separated from our attitude to society. Even if religion arose as the spiritual structure of a particular human society, it is arguable whether it has remained the strongest social molding force throughout history, or whether society, once formed, develops new spiritual structures and adapts them to its particular level of knowledge. Nowadays, the individual seems to be able to choose the spiritual framework of his thoughts and actions quite freely, and this freedom reflects the fact that the boundaries between the various cultures and societies are beginning to become more fluid. But even when an individual tries to attain the greatest possible degree of independence, he will still be swayed by the existing spiritual structures—consciously or unconsciously. For he, too, must be able to speak of life and death and the human condition to other members of the society in which he has chosen to live; he must educate his children according to the norms of that society, fit into its life. Epistemological sophistries cannot possibly help him attain these ends. Here, too, the relationship between critical thought about the spiritual content of a given religion and action based on the deliberate acceptance of that content is complementary. And such acceptance, if consciously arrived at, fills the individual with strength of purpose, helps him to overcome doubts and, if he has to suffer, provides him with the kind of solace that only a sense of being sheltered under an all-embracing roof can grant. In that sense, religion helps to make social life more harmonious; its most important task is to remind us, in the language of pictures and parables, of the wider framework within which our life is set."

"You keep referring to the individual's free choice," I said, "and you compare it with the freedom with which the atomic physicist can arrange his experiments in this way or that. Now the classical physicist had no such freedom. Does that mean that the special features of modern physics have a more direct bearing on the problem of the freedom of the will? As you know, the fact that atomic processes cannot be fully determined is often used as an argument in favor of free will and divine intervention."

"I am convinced that this whole attitude is based on a simple misunderstanding, or rather on the confusion of questions, which, as far as I can see, impinge on distinct though complementary ways of looking at things. If we speak of free will, we refer to a situation in which we have to make decisions. This situation and the one in which we analyze the motives of our actions or even the one in which we study physiological processes, for instance the electrochemical processes in our brain, are mutually exclusive. In other words, they are complementary, so that the question whether natural laws determine events completely or only statistically has no direct bearing on the question of free will. Naturally, our different ways of looking at things must fit together in the long run, i.e., we must be able to recognize them as noncontradictory parts of the same reality, though we cannot yet tell precisely how. When we speak of divine intervention, we quite obviously do not refer to the scientific determination of an event, but to the meaningful connection between this event and others or human thought. Now this intellectual connection is as much a part of reality as scientific causality; it would be much too crude a simplification if we ascribed it exclusively to the subjective side of reality. Once again we can learn from the analogous situation in natural science. There are well-known biological relations that we do not describe causally, but rather finalistically, that is, with respect of their ends. We have only to think of the healing process in an injured organism. The finalistic interpretation has a characteristically complementary relationship to the one based on physico-chemical or atomic laws; that is, in the one case we ask whether the process leads to the desired end, the restoration of normal conditions in the organism; in the other case we ask about the causal chain determining the molecular processes. The two descriptions are mutually exclusive, but not necessarily contradictory. We have good reason to assume that quantum-mechanical laws can be proved valid in a living organism just as they can in dead matter. For all that, a finalistic description is just as valid. I believe that if the development of atomic physics has taught us anything, it is that we must learn to think more subtly than in the past."

"We always come back to the epistemological side of religion," I objected. "But Dirac's attack on religion was aimed chiefly at its ethical side. Dirac disapproves quite particularly of the dishonesty and self-deception that are far too often coupled to religious thought. But in his abhorrence he has become a fanatic defender of rationalism, and I have the feeling that rationalism is not enough."

"I think Dirac did well," Niels said, "to warn you so forcefully against the dangers of self-deception and inner contradictions; but Wolfgang was equally right when he jokingly drew Dirac's attention to the extraordinary difficulty of escaping this danger entirely." Niels closed the conversation with one of those stories he liked to tell on such occasions: "One of our neighbors in Tisvilde once fixed a horseshoe over the door to his house. When a mutual acquaintance asked him, 'But are you really superstitious? Do you honestly believe that this horseshoe will bring you luck?' he replied, 'Of course not; but they say it helps even if you don't believe in it.' "

8

Atomic Physics
and Pragmatism (1929)

To those of us who participated in the development of atomic theory, the five years following the Solvay Congress in Brussels looked so wonderful that we often spoke of them as the golden age of atomic physics. The great obstacles that had occupied all our efforts in the preceding years had been cleared out of the way; the gate to that entirely new field—the quantum mechanics of the atomic shell—stood wide-open, and fresh fruits seemed ready for the plucking. Where purely empirical rules or vague concepts had had to serve as substitutes for real understanding— for instance, of ferromagnetic phenomena and of chemical bonds in the physics of solids—the new methods brought absolute clarity. Moreover, it seemed very much as if the new physics was in many respects greatly superior to the old even on the philosophical plane; that, in ways that had to be investigated more closely, it was much broader and richer.

In the late autumn of 1927 when I was offered a professorship by the universities of both Leipzig and Zurich, I decided for the former, where I would be working with the brilliant experimental physicist, Peter Debye. Though my first seminar on atomic theory was attended by just one student, I was convinced that I would eventually make many fresh converts to the new atomic physics.

Before taking over my new post, I was granted a year's leave of absence to go on a lecture tour to the United States. And so, in February 1929 during a particularly cold spell, I boarded a ship in Bremerhaven for New York. It took us two whole days to get out of the harbor: the shipping channel was blocked with thick ice, and, once outside, we were tossed about by the most violent storms that I had ever experienced at sea. Then, after fifteen rough days the coast of Long Island and, later at dusk, the famous skyline of New York finally rose up to bid us welcome.

The New World cast its spell on me right from the start. The carefree attitude of the young, their straightforward warmth and hospitality, their gay optimism—all this made me feel as if a great weight had been lifted from my shoulders. Interest in the new atomic theory was keen, and since I had been invited by a fairly large number of universities in many parts of the country, I became acquainted with many different aspects of American life. Wherever I stayed for more than a few days, I struck up new acquaintanceships that started with tennis, boating or sailing parties and quite often ended in long discussions of recent developments in atomic physics. I quite particularly remember a conversation with my tennis partner, Barton Hoag, a young experimental physicist from Chicago, who invited me to join him on a fishing trip to the remote northern lakes.

I told him of a strange feeling I had acquired during this lecture tour: while Europeans were generally averse and often overtly hostile to the abstract, nonrepresentational aspects of the new atomic theory, to the wave-corpuscle duality and the purely statistical character of natural laws, most American physicists seemed prepared to accept the novel approach without too many reservations. I asked Barton how he explained the difference, and this is what he said:

"You Europeans, and particularly you Germans, are inclined to treat such new ideas as matters of principle. We take a much simpler view. Newtonian physics used to provide an accurate enough description of the observed facts. Then we became acquainted with electromagnetic phenomena, and found that Newtonian mechanics was no longer adequate, but that Maxwell's equations did the trick. Finally, the study of atomic processes taught us that neither classical mechanics nor electrodynamics could account for the experimental evidence. And so physicists had willy-nilly to go beyond the old laws or equations. The result was quantum mechanics. Basically, physicists, even the theorists among them, behave just like the engineer building a new bridge. He notices that the old formulae he has been using in the past do not quite fit the new construction. He must allow for wind pressure, for metal fatigue, for temperature variations and the like, all of which he now builds into the old formulae. The result is a more reliable blueprint, and everyone is happy about it. But the basic engineering principles have remained unchanged. The same seems to be true of modern physics. Perhaps you make the mistake of treating the laws of nature as absolutes, and you are therefore surprised when they have to be changed. To my mind, even the term 'natural law' is a glorification or sanctification of what is basically nothing but a practical prescription for dealing with nature in a particular domain. I believe that once all absolutist claims are dropped, the difficulties will disappear by themselves."

"Then you are not at all surprised," I asked, "that an electron should appear as a particle on one occasion and as a wave on another? As far as you are concerned, the whole thing is merely an extension of the older physics, perhaps in unexpected form?"

"Oh, no, I am surprised; but, after all, I can see that it happens in nature, and that's that. If there are structures that look like a wave on one occasion and like a particle on the next, then we must obviously come up with new concepts. Perhaps one ought to call such structures 'wavicles,' and quantum mechanics the mathematical description of their behavior."

"No, that solution is a bit too simple for me. After all, we are not dealing with a special property of electrons, but with a property of all matter and of all radiation. Whether we take electrons, light quanta, benzol molecules or stones, we shall always come up against these two characteristics, the corpuscular and the undular. In other words, the statistical features of natural laws are ubiquitous and a matter of principle. It's just that these quantum-mechanical features are far more obvious in atomic structures than in the objects of daily experience."

"Very well, then, you have simply introduced minor changes into the Newtonian and Maxwellian laws, changes that are more obvious to the observer of atomic phenomena, and imperceptible to people working in more down-to-earth fields. In either case, the changes represent improvements, and no doubt quantum mechanics will also be further improved in the future so as to account for other phenomena that still elude our grasp. In the meantime, quantum mechanics strikes me as a correct procedure in the atomic realm, one that has obviously proved its worth."

I found Barton's whole approach somewhat unsatisfactory, but I realized I would have to be much more explicit if I were to make him see why. Hence I said rather pointedly: "I think Newtonian mechanics cannot be improved in any way, for inasmuch as we can describe a particular phenomenon with the concepts of Newtonian physics—namely, position, velocity, acceleration, mass, force, etc.—Newton's laws hold quite rigorously, and nothing in this will be changed for the next hundred thousand years. More precisely, I ought perhaps to say: Newton's laws are valid to that degree of accuracy to which the phenomena concerned can be described by these concepts. The fact that this accuracy has limits, was of course well known even to classical physicists, none of whom ever claimed he could measure to any desired degree of accuracy. The fact, however, that the accuracy of measurements is limited in principle, i.e., by uncertainty relations, is something quite new, something we first encountered in the atomic field. But for the moment we need not enter into this subject. For the purposes of our discussion, it is enough to assert that, inasmuch as it is possible to make accurate measurements of this kind at all, Newtonian mechanics is fully valid now and will remain so in the future."

"I'm afraid I cannot follow you here," Barton replied. "Isn't the mechanics of relativity theory an improvement on Newtonian mechanics? Even if we leave the uncertainty principle out of the discussion?"

"We may leave the uncertainty principle out of the discussion," I tried to point out, "but not the fact that we are dealing with a different space-time structure, and quite particularly with a different relationship between time and space. As long as we talk of absolute time, i.e., of time that is apparently independent of the observer's place and state of motion, as long as we deal with rigid or practically rigid bodies of given volume, then Newton's laws do hold. But as soon as we come to processes involving very great velocities and at the same time try to make very precise measurements, we discover that Newtonian mechanics no longer provides an adequate description. We shall find, for instance, that the clock of a moving observer appears to go more slowly than that of an observer at rest, etc., and then we must have recourse to relativistic mechanics."

"But why aren't you prepared to call relativistic mechanics an improvement on Newtonian mechanics?"

"I only objected to the term 'improvement' because it may lead to misunderstandings, but once that danger is removed I have no further objections. The misunderstanding hinges precisely on your idea that progress in physics is comparable to improvements in the engineering field. To my mind, it is fundamentally wrong to compare the basic changes involved in the transition from Newtonian mechanics to relativistic or quantum mechanics with the improvements of the engineer, which, after all, do not call for modifications of his basic concepts. For him, all technical terms retain their old significance; at most, the formulae are adjusted or corrected so as to cover previously neglected factors. Changes of this type, however, would make no sense at all in Newtonian mechanics. There are no experiments to force them upon us. And this is precisely why we can grant that Newtonian physics has an absolute validity: in its particular sphere of application it cannot be improved by small changes. However, there are areas of experience in which we can no longer manage with the conceptual system of Newtonian mechanics. For such areas we need entirely new conceptual structures, for instance, those introduced by relativity theory or quantum mechanics. Still, Newtonian physics constitutes a closed system in the sense that the physical equipment of the engineer can never hope to be. It is thanks to this coherence that there can be no minor improvements. All we can do is to adopt an entirely new conceptual system, in which the old system is contained as a limiting case."

"How can you know," Barton asked, "whether a particular realm of physics is closed in the sense you claim Newtonian mechanics is? Which criteria distinguish closed realms from those that are still open, and which are the closed realms of present-day physics?"

"The most important criterion for a closed system is probably the presence of a precisely formulated and self-consistent set of axioms governing the concepts and logical relations of the system. To what extent an axiomatic system corresponds to reality can only be decided empirically, and we can only call it a 'theory' if it represents large realms of experience. On that basis, I would distinguish four closed realms in physics: Newtonian mechanics, statistical thermodynamics, special relativity theory together with Maxwell's electrodynamics and, finally, modern quantum mechanics. For each of these realms there is a precisely formulated system of concepts and axioms, whose propositions are strictly valid within the particular realm of experience they describe. I think it is too early to count the general theory of relativity among the closed realms, because its axiom system is still unclear and its application to cosmological problems still admits of various solutions. For the time being, we must therefore treat it as one of the open theories, one that is still full of unanswered questions."

Barton seemed fairly satisfied with my reply, but wanted to know more about my motives for introducing the doctrine of closed systems. "Why do you lay so much stress on the fact that the transition from one realm to another, for instance from Newtonian physics to quantum theory, is not continuous but discrete? You are right, of course, in saying that new concepts are introduced, and that different questions are asked in each new realm. But why is that so important? After all, what matters is the progress of science, the opening up of ever wider realms of nature. Whether this progress is continuous or occurs by steps seems a matter of complete indifference to me."

"No, it is anything but that. Your idea of continuous progress as we know it from engineering would weaken, or rather soften, physics to such an extent that we could hardly continue to call it an exact science. If we wanted to work at physics in this purely pragmatic way, we would have to keep picking on what partial realms happened to be experimentally accessible, and then try to represent the phenomena in them by approximations. If the results turned out to be too inaccurate, we could always add fresh corrections. But we should have to give up asking about the wider connections, and there would be very little chance of our advancing to the very simple relations which, to mention just one example, distinguish Newtonian mechanics from Ptolemy's astronomy. In other words, we would lose the most important truth criterion of physics, namely, the ultimate simplicity of all physical laws. You may, of course, object that this insistence on simplicity is nothing but a hidden thirst for the absolute, devoid of the least logical justification. Why should physical laws be simple, why should wide realms of experience be susceptible to simple representation? The answer lies in the history of physics. You will admit that each of the four closed realms I have mentioned has a very simple axiom system, capable of embracing a very wide set of relations. Only in such axiom systems are we entitled to speak of 'physical laws'; without them physics would never have attained the noble rank of an exact science.

"This simplicity has still another consequence, one that affects our relationship to physical laws. But this whole subject is extremely difficult to put into words. If, as we must always do as a first step in theoretical physics, we combine the results of experiments and formulae and arrive at a phenomenological description of the processes involved, we gain the impression that we have invented the formulae ourselves. If, however, we chance upon one of those very simple, wide relationships that must later be incorporated into the axiom system, then things look quite different. Then we are quite suddenly brought face to face with a relationship that has always existed, and that was quite obviously not invented by us or by anyone else. Such relationships are probably the real content of our science. Only when one has fully assimilated the fact of their existence can one really claim to have grasped physics."

Barton reflected. He did not contradict me, but I had the distinct impression that my way of thinking was rather alien to him.

Luckily our weekend was not entirely filled with such intense conversations. We spent our first night in a small hut on the shores of a secluded lake, in the midst of what looked like an endless stretch of water and forests. Next morning, an Indian guide took us out fishing, and so well did he know the locality that, within the hour, we had bagged eight unusually large pike which provided us and our guide's family with a most satisfying dinner. Next day, we tried again, but this time without the Indian. Weather and wind conditions were much the same, and we made straight for the same spot in the lake. But though we stayed out all day, we came back empty-handed. "These fish," Barton said, "are just like atoms. If you aren't fully familiar with their intimate habits and reactions, you have little chance of ever catching them."

Toward the end of my stay in America, I made arrangements to return home with Paul Dirac. We would meet in Yellowstone Park, go walking for a few days, then sail across the Pacific to Japan and return to Europe via Asia. Our meeting place was the hotel in front of Old Faithful, and since I got there a day earlier, I did some mountaineering on my own. Only on the way up did I realize that these mountains, unlike the Alps, are rarely explored by man: there were neither roads nor footpaths, neither signs nor markings, and no help could be expected in an emergency. On the way up I wasted a great deal of time on a roundabout route, and during the descent I became so tired that I sat down in the grass and fell asleep at once. I was awakened by a bear licking my face. I got such a shock that I started off at once, but it was not until dusk that I was able to find my way back to the hotel.

In my letter to Paul Dirac I had mentioned that we might perhaps look at some of the other geysers, and that with luck we might see one or two in action. It was characteristic of Paul's careful and systematic ways that, when we met, he had already worked out a precise itinerary in which he had not only marked the times of activity of these natural fountains, but had mapped out precise routes that would bring us, in the course of one afternoon, to the greatest possible number of geysers just in time to watch them spring into action.

Physics was discussed chiefly during our long sea voyage from San Francisco to Yokohama via Hawaii. Although I was happy to join in the many games of table tennis or shuffleboard, there were hours in which, comfortably ensconced in a deck chair, I could watch schools of dolphins bustling about the ship or swarms of flying fish making a quick getaway. Since Paul usually took a deck chair next to mine, we could speak at length about our respective experiences in America and our ideas about the future of atomic physics. The readiness of American physicists to accept even the abstract, nonrepresentational features of the new atomic physics surprised Paul much less than it did me. Like Barton, he probably felt that the development of our science was a more or less continuous process, in which clarifying the conceptual structures that emerged at any particular stage of development mattered less than discovering the quickest path to the next stage. For once you use the pragmatic approach, you are bound to consider the progress of science as a continuous and never-ending process of thought adaptation to the growing body of experimental knowledge. What matters, therefore, is not the prevailing interpretation, but the method of adaptation.

Like me, Paul was convinced that simple physical laws would finally emerge, or, as I would put it, come to light, during this process. But, methodologically, his starting points were particular problems, not the wider relationship. When he described his approach, I often had the feeling that he looked upon scientific research much as some mountaineers look upon a tough climb. All that matters is to get over the next three yards. If you do that long enough, you are bound to reach the top. To keep thinking of the whole climb with all its innumerable difficulties only leads to discouragement. And, in any case, you only grasp the true problems when you reach the most difficult ledges. I myself took a different view. My first step—to stick to the mountaineering simile—was a decision about the climb as a whole. For I was convinced that once one has found the correct route, then and then only can the individual obstacles be surmounted. The whole comparison was false, however, because, in the case of a rock ledge, you can never tell in advance what lies behind and above, whereas in science the basic relations have to be simple; nature, I was certain, is made to be understood, or, rather, our thought is made to understand nature. As Robert had put it so well during our walk around Lake Starnberg: the same organizing forces that have shaped nature in all her forms are also responsible for the structure of our minds.

Paul and I spoke a great deal about this methodological question, and about our hopes with respect to future developments: whereas Paul argued that one can never solve more than one difficulty at a time, I contended that one can never overcome an isolated difficulty, but must always surmount several at once. Paul probably wished to say no more than that anyone trying to grapple with more than one problem at a time was guilty of arrogance. For he knew how hard one has to fight for every step in a realm as remote from daily experience as atomic physics. I, for my part, merely wished to point out that the genuine solution of a difficult problem is neither more nor less than a glimpse of the wider context, a glimpse that helps us to clear away other difficulties as well, including many whose existence we do not even suspect. And so both our formulations contained a large grain of truth, and Paul and I could console ourselves with an oft-repeated dictum of Niels Bohr: "The opposite of a correct statement is a false statement. But the opposite of a profound truth may well be another profound truth."

9

The Relationship between Biology, Physics and Chemistry (1930–1932)

From America and Japan, I returned to a wide round of duties in Leipzig. I had to give lectures, prepare tests, participate in faculty meetings and examinations, help to modernize our tiny Institute of Theoretical Physics and introduce young physicists to quantum theory. So much academic variety was quite new to me, and I thoroughly enjoyed it. Still, my links with the Copenhagen circle around Niels Bohr had become so important over the years that I made a point of spending a few weeks of most of my holidays in Denmark. Some of the most vital discussions were held in Bohr's country house in Tisvilde or on a sailing boat which he and some of his friends kept on Langelinie in Copenhagen Harbor, from which we would often set out far into the Baltic.

The country house was situated in Northern Zealand, a mile or so from the beach and at the edge of a large forest. I knew it well from our first walk. To reach the popular beach we had to take a sandy forest path; its straightness suggested that we were in a man-made plantation, serving as a screen against storms and dune migrations. When Niels' children were still small, he used to keep a horse and cart, and I always thought it a special honor if I was allowed to drive one of the youngsters through the woods.

In the evenings, we would often sit around the open fire, taking great pains to keep it alight. When the doors of the living room were closed, the chimney would smoke a good deal, and we had to leave at least one door open. The result was a mighty draft, and Niels, who loved paradoxical formulations, claimed that the fireplace had been especially put in as a cooling device. But hot or cold, the area round the fireplace was highly popular, and whenever other physicists came to visit us from Copenhagen, keen conversations on the problems that interested all of us would take place before it. I remember one evening particularly well. I think that Kramers and Oskar Klein were among those present, and, as so often, the discussion revolved around Einstein's refusal to accept the statistical character of the new quantum mechanics.

"Isn't it odd," Oskar Klein said, "that Einstein should have such great difficulties in accepting the role of chance in atomic physics? He knows more about statistical thermodynamics than most other physicists, and has himself produced a convincing statistical derivation of Planck's radiation law. Yet he still rejects quantum mechanics, simply because chance plays a fundamental part in it."

"It is precisely this fundamental aspect that upsets him," I tried to point out. "In a pot of water, we cannot possibly hope to tell how each water molecule moves. Hence no one should be surprised that physicists are applying statistics to it, making use of probability, much as life insurance companies must make actuarial computations of their clients' life expectancies. But we used to assume that, in principle at least, it was possible to describe the motion of every molecule in accordance with the laws of Newtonian mechanics. In other words, nature was thought to have at any given moment an objective state from which one could deduce its state during the next moment. But this is no longer so in quantum mechanics. Here we cannot make observations without disturbing the phenomena—the quantum effects we introduce with our observation automatically introduce a degree of uncertainty into the phenomenon to be observed. This Einstein refuses to accept, although he knows the facts perfectly well. He thinks that our interpretation cannot possibly be complete, and hopes that the discovery of fresh data will help to close what he thinks are open gaps in our knowledge. But that is an idle hope."

"I don't entirely agree with you," Niels said. "There is, of course, a basic difference between classical thermodynamics and quantum mechanics, but you have exaggerated its importance. In any case, I find all such assertions as 'observation introduces uncertainty into the phenomenon' inaccurate and misleading. Nature has taught us that the word 'phenomenon' cannot be applied to atomic processes unless we also specify what experimental arrangement or what observational instruments are involved. If a particular experimental setup has been defined and a particular observation follows, then we can admittedly speak of a phenomenon, but not of its disturbance by observation. And though the results of different observations can no longer be correlated as directly as they could in classical physics, this does not mean that the phenomena have been disturbed by observation; it simply means that we cannot objectify the observational results in the manner of classical physics or everyday experience. Different observational situations—by that I mean the over-all experimental setup, the readings, etc.—are often complementary, i.e., they are mutually exclusive, cannot be obtained simultaneously, and their results cannot be correlated without further ado. Hence I cannot see any very fundamental difference between quantum mechanics and thermodynamics. An observational situation involving a temperature reading also has an exclusive relationship with one in which the coordinates and velocities of all the participating particles can be determined. After all, the very concept of temperature may be defined as the degree of uncertainty about the behavior of fractions of the system characteristic of what we call 'canonical distribution.' Or, to put it less academically: if a system consisting of many particles exchanges heat steadily with the environment or with other macrosystems, then the energy of each particle will fluctuate continuously and so will that of the the entire system. However, the mean values obtained from a large number of particles over long periods of time correspond very precisely to the mean values of this normal or canonical distribution. All this you can read in Gibbs. And temperature, after all, can only be defined by energy exchanges. It follows that a precise determination of temperature is incompatible with a precise determination of the positions and veloc ities of the molecules."

"But doesn't that mean," I asked, "that temperature is not an objective property? We have always thought that the statement 'The tea in this pot has a temperature of 70 degrees' refers to an objective fact; that everyone who measures the temperature in that particular teapot will get a reading of 70 degrees, regardless of how he performs the measurement. On the other hand, if temperature merely defines the degree of one's knowledge or ignorance of the molecular motions in the tea, it follows that different observers may obtain different temperature readings even if the real state of the system is identical; after all, different observers can have different levels of knowledge."

"No, you are quite wrong," Niels broke in. "The very word 'temperature' refers to an observational situation involving energy exchanges, say, between the tea and the thermometer, and this quite irrespective of the other properties of the thermometer. A thermometer is only a real thermometer if the molecular motions in the system to be measured, in our case the tea, and in the thermometer itself reflect the canonical distribution with the required degree of accuracy. If that is the case, all thermometers will give the same readings, and to that extent temperature is an objective quality. It all goes to show once again how problematical the concepts 'objective' and 'subjective,' which we normally use so glibly, really are."

Kramers seemed unhappy about this definition. "From the way you speak of processes inside the teapot," he said to Niels, "one might think that you were asserting some sort of uncertainty relation between the temperature and the energy inside. But surely you can't be thinking of applying such ideas in classical physics?"

"To a certain extent, I can," Niels replied. "Let us look at the properties of an individual hydrogen atom inside the teapot. Its temperature, if we can talk about it at all, is surely as high as that of the rest of the tea, in our case 70 degrees, because it exchanges heat with all the other tea molecules. Its energy, however, must fluctuate and this precisely because it exchanges heat; hence we can only define a probability curve for its energy. If, conversely, we had measured the energy of the hydrogen atom and not the temperature of the tea, then we could not deduce the latter unequivocally from the former; once again we could only draw a probability curve—for the temperature. The relative breadth of this curve, in other words the lack of certainty as to the precise temperature or energy values, is relatively large in so small an object as a hydrogen atom, and hence significant. In a much larger object, for instance a small quantity of tea within the pot, it becomes considerably smaller and can therefore be neglected."

"In the old thermodynamics as we teach it to our students," Kramers objected, "energy and temperature are attributed to an object simultaneously. Not a word is said about indeterminacy or uncertainty. How can you possibly reconcile that with your views?"

"The old thermodynamics," Niels replied, "is to statistical thermodynamics what classical mechanics is to quantum mechanics. With large objects we do not commit significant errors if we attribute certain values to their temperature and energy simultaneously, just as we can attribute simultaneous values to their position and velocity. With very small objects things are quite different. In thermodynamics, we used to say that the latter are endowed with energy but not with temperature. But this strikes me as a mistaken idea, if only because we don't know where to draw the line between small and large objects."

All of us now realized why Niels laid far less stress on the fundamental difference between the statistical laws of thermodynamics and those of quantum mechanics than Einstein. Niels felt that complementarity was a central feature of all attempts to describe nature, a feature inherent in, though insufficiently brought out by, statistical thermodynamics, especially in the form Gibbs had given it; Einstein, on the other hand, remained steeped in the conceptual world of Newtonian mechanics or of Maxwellian field theory and completely failed to notice these complementary features.

The discussion now turned to possible further applications of complementarity, and Niels mentioned that it might well help in distinguishing biological processes from purely physical and chemical ones. This subject was discussed at greater length during one of our sailing trips, of which I shall now say a few words.

The captain of our boat was Niels Bjerrum, a chemical physicist at the University of Copenhagen, a man who combined the wry humor of the old salt with a thorough knowledge of navigation. From the outset, his attractive personality filled me with so much confidence that I would have followed his orders in any situation. Among the crew was the surgeon, Chievitz, who liked to make ironical remarks about everything that happened on board, and who often chose the captain as the special butt of his friendly gibes. Bjerrum gave as good as he got, and it was great fun to listen to their exchanges. On this particular trip, there were also two other crew members whose names I cannot remember.

At the end of each summer, the yacht *Chita* had to be taken from Copenhagen to Svendborg on the island of Fyn, where she was put in dry dock and repaired. The trip could not be managed in one day even with a following wind, and we made our arrangements accordingly. We set out from Copenhagen very early one morning, with a fresh northwesterly and under a bright sky. We quickly passed the southern tip of the island of Amager and sailed into Köge Bay on a southwesterly course. After a few hours the steep Stevns Klint came into view, and as soon as we had passed it, the wind dropped. We were almost completely becalmed, and after one or two hours of this we began to grow impatient. We had earlier been discussing disastrous North Pole expeditions, and Chievitz now said to Bjerrum: "If the wind doesn't freshen up soon, we'll run out of supplies and then we'll have to draw lots as to who'll be eating whom first." Bjerrum handed Chievitz a bottle of beer and said: "I didn't realize that you would be needing spiritual solace quite so soon. Let's hope this bottle will keep you going through the next hour."

All of a sudden, the sky became clouded over, and the first drops of rain began to fall on the deck. We had to put on our oilskins. When we entered the narrow channel between the islands of Zealand and Möen, the wind was nearly up to gale force and the rain had turned into a downpour. We had to tack so hard in the small fairway that after an hour or so all of us were close to exhaustion. My hands had swelled up from the unaccustomed work with the ropes, and Chievitz murmured: "What a pity our captain didn't find us an even smaller fairway. Still, we are sailing for pleasure, so we mustn't complain too much." Niels lent a hand with all the difficult maneuvers, and I was amazed at how much physical strength he still had. It was dusk when, at long last, we reached Storström Strait, a broad stretch of water between Zealand and Falster. Since we were making northwest now, and the rain had stopped, we could relax and let the wind carry us forward. Sailing by the compass in complete darkness, we began to chat, occasionally taking our bearings from some distant beacons. Some of the crew had taken to the small cabin, resting or sleeping after their hard labor. Chievitz was at the wheel, Niels stood next to him glancing at the compass, and I posted myself forward, looking out for the lights of passing ships. "It's all very well watching for lights," Chievitz piped up, "but what if we should meet a stray whale? It'll neither have red lights to port nor green lights to starboard and we might easily hit it. Heisenberg, can you see any whales?"

"I can't see anything but," I replied, "though some of them may turn out to be nothing more dangerous than big waves."

"Let's hope so. But what if we *should* hit a whale? Our boat and the whale would both get holed. But there is this fundamental difference between the two: the hole in the whale would heal by itself, while our boat would remain a wreck. Particularly if it were lying at the bottom of the sea. Otherwise, of course, we can always get it repaired."

Niels now joined in. "The difference between living and dead matter is not nearly so simple. True, in the whale we can see the workings of a formative force, if that's what you want to call it, which ensures that the injured part becomes whole again. Naturally, the whale itself knows nothing about this formative force. No doubt it is some unexplained part of its biological heritage. But the ship, too, is not a completely dead object. It behaves toward men much as the web behaves toward the spider or the nest toward the bird. Its formative force emanates from man, and the process of repair is somehow analogous to the process of healing. For if there were no living being—in this case man—to determine its shape, the boat would never get repaired either. Of course, the fact that, in man, this formative force involves consciousness does constitute an important difference."

"By formative force," I asked, "do you refer to something quite outside the realm of physics and chemistry, or do you imagine this force can somehow express itself in the position of atoms, in their mutual interactions, in resonance effects or the like?"

"We should probably have to start from the fact that an

organism has the kind of wholeness that a system built up of a host of atomic bricks—the kind considered by classical physics—could never have," Niels said. "But then we have moved on to quantum mechanics. Hence we may be tempted to compare those integral structures quantum theory can represent in mathematical terms, for instance the stationary states of atoms and molecules, with those resulting from biological processes. But here, too, there are a number of fundamental differences. The integral structures of atomic physics—atoms, molecules, crystals—are all of them static structures; they consist of a certain number of elementary particles, atomic nuclei and electrons, and they do not change with time unless they are disturbed from the outside. If that happens, then they will admittedly react, but if the disturbance is not too great or persistent, they will eventually return to their original state. Organisms, by contrast, are anything but static structures. The ancients used to compare living beings to flames because they thought of both as forms through which matter streams. It is quite unlikely that we shall ever be able to determine by measurements which particular atoms belong to a living being and which do not. The question must therefore be put as follows: Can quantum mechanics explain nature's tendency to form structures through which matter with fixed chemical properties can stream for a limited time?"

"The physician," Chievitz interjected, "does not have to bother about the answer to that question. He assumes that the organism has a tendency to return to normal conditions after a disturbance, and he is also convinced that the processes involved are causally determined; in other words, that a mechanical or chemical intervention leads to the very effects predicted by physics and chemistry. The fact that the biological and physical approaches may be incompatible does not even occur to most physicians."

"But here we have a typical case of two complementary ways of looking at things," Niels pointed out. "We can, first, describe an organism with concepts men have developed through contact with living beings over the millennia. In that case, we speak of 'living,' 'organic function,' 'metabolism,' 'breathing,' 'healing,' etc. Or else we can inquire into causal processes. Then we use the language of physics and chemistry, study chemical or electrical processes, for instance in nerve conduction, and assume, apparently with great success, that the laws of physics and chemistry, or more generally the laws of quantum theory, are fully applicable to living organisms. These two ways of looking at things are contradictory. For in the first case we assume that an event is determined by the purpose it serves, by its goal. In the second case we believe that an event is determined by its immediate predecessor. It seems most unlikely that both approaches should have led to the same result by pure chance. In fact, they complement each other, and, as we have long since realized, both are correct precisely because there is such a thing as life. Biology thus has no need to ask which of the two viewpoints is the more correct, but only how nature managed to arrange things so that the two should fit together."

"In other words," I cut in, "you do not believe that, over and above the forces and mutual effects known in atomic physics, there exists a special life force—for instance, the kind stipulated by the vitalists—a force responsible for the special behavior of living organisms, in our case the whale's recovery from his wounds? You would rather take the view, wouldn't you, that the characteristic biological laws, for which no analogy can be found in inorganic matter, result from what you have just described as a complementary situation?"

"Yes, I would agree with that. One could also say that the two ways of looking at things refer to complementary observational situations. In principle, we could probably measure the position of every atom in a cell, though hardly without killing the living cell in the process. What we would know in the end would be the arrangement of the atoms in a dead cell, not a living one. If we now used quantum mechanics to determine the subsequent behavior of these atomic arrangements, we would find that the cell decays, breaks up or whatever else you care to call it. If, conversely, we want to keep the cell alive and hence allow no more than very cautious observations of the atomic structure, then our conclusions will still be correct, but they will not enable us to tell whether the cell survives or decays."

"I can quite see why you should distinguish biological from physical and chemical laws by means of complementarity," I said. "But your remarks still leave us a choice between two interpretations that many scientists consider as different as chalk and cheese. Let us imagine that one day biology will have become as completely fused with physics and chemistry as physics and chemistry have become fused in quantum mechanics. Do you believe that the natural laws of this new science will simply be the laws of quantum mechanics with some biological concepts superimposed, just as such statistical concepts as temperature and entropy have been superimposed on the laws of Newtonian mechanics? Or do you, rather, believe that this unified science will be governed by broader natural laws of which quantum mechanics is only a limiting case, much as Newtonian mechanics may be considered a limiting case of quantum mechanics? In favor of the first assumption we can argue that we must, in any case, add the concept of evolution, i.e., of selection in geological time, to quantum mechanics, if we are to explain the profusion of organisms. There is no reason why the addition of this element should cause us any fundamental difficulties. Living organisms would simply be those forms nature has tried out on earth in the framework of quantum-mechanical laws over several thousands of millions of years. But there are also good arguments for the second view. For instance, we might say that nothing in quantum theory so far points to a tendency to produce forms capable of maintaining fixed chemical properties, despite continuous exchanges of matter. I cannot tell which of the two arguments has the greater weight. Can you?"

"First of all, I fail to see," Niels said, "why a choice between these viewpoints should be a matter of such importance to science in its present state. What is important is to find a proper place for biology in a world so dominated by physical and chemical laws. This, the concept of the complementarity of observational situations enables us to do. The eventual addition of biological concepts to quantum mechanics is a foregone conclusion. However, we cannot yet tell whether the addition of the former will necessarily call for an extension of the latter. Perhaps the wealth of mathematical forms hidden in quantum mechanics is large enough to embrace biological forms. As long as biological research sees no reason for the extension of quantum physics, we ourselves should certainly not insist. In science it is always the best policy to be as conservative as possible, and to make extensions only if the observations would otherwise remain inexplicable."

"But there are biologists who believe that this is already the case," I remarked. "They think that Darwinism in its present form—chance mutation and selection—cannot possibly account for all the different organic forms we find on earth. True, the layman has no difficulty when the biologist explains the existence of accidental mutations, tells him that the genetic stock of a given species may be subjected to sudden changes in this or that direction, and that environmental factors favor the propagation of some species and suppress that of others. Darwin's explanation, that the whole thing is a selective process, that only the fit survive, is also readily believed, though one may perhaps wonder whether this statement is a scientific assertion or simply a definition of the word 'fit.' We call all those species 'fit' or 'viable' which prosper under the given circumstances. But even if we agree that selection leads to the emergence of particularly fit or viable species, it is very difficult to believe that such complicated organs as, for instance, the human eye were built up quite gradually as the result of purely accidental changes. Many biologists obviously take the view that this is precisely what did happen, and they will also tell you what particular steps in the course of geological history could have led to the final result, the eye. Others are more skeptical. I have been told about a conversation between von Neumann, the mathematician, and a biologist. The biologist was a convinced neo-Darwinist; von Neumann was skeptical. He led the biologist to the window of his study and said: 'Can you see the beautiful white villa over there on the hill? It arose by pure chance. It took millions of years for the hill to be formed; trees grew, decayed and grew again, and then the wind covered the top of the hill with sand, stones were probably deposited on it by a volcanic process, and accident decreed that they should come to lie on top of one another. And so it went on. I know, of course, that accidental processes through the aeons generally produce quite different results. But on just this one occasion they led to the appearance of this country house, and people moved in and live there at this very moment.' Needless to say, the biologist disliked this line of reasoning. But though von Neumann is no biologist, and though I myself can't judge who was right, I take it that even among biologists there is some hesitation as to whether Darwinian selection provides an adequate explanation of the existence of the most complicated organisms."

"The whole thing is probably a question of the correct time scale," Niels suggested. "Darwinian theory in its present form makes two independent assertions. On the one hand, it states that, through the process of heredity, nature tests ever new living forms, rejecting the great majority and preserving a few suitable ones. This seems to be empirically correct. But there is also the second assertion: that the new forms originate through purely accidental disturbances of the gene structure. This claim is much more questionable, even though we can hardly conceive of an alternative. Von Neumann's argument was, of course, designed to show that, though almost anything can arise by chance in the long run, the probability of this happening in the time we know nature has taken to produce higher organisms is absurdly small. Physical and astrophysical studies tell us that no more than a few thousand million years have passed since the appearance of the most primitive living beings on earth. Now, whether or not accidental mutations and selection are sufficient to produce the most complicated and highly developed organisms during this interval will depend on the time needed to develop a new biological species. I suspect that we know much too little about this factor to give a reliable answer. Hence we had best suspend judgment for the time being."

"Another argument," I continued, "that is occasionally brought up in favor of an extension of quantum theory is the existence of human consciousness. There can be no doubt that 'consciousness' does not occur in physics and chemistry, and I cannot see how it could possibly result from quantum mechanics. Yet any science that deals with living organisms must needs cover the phenomenon of consciousness, because consciousness, too, is part of reality."

"This argument," Niels said, "looks highly convincing at first sight. We can admittedly find nothing in physics or chemistry that has even a remote bearing on consciousness. Yet all of us know that there is such a thing as consciousness, simply because we have it ourselves. Hence consciousness must be part of nature, or, more generally, of reality, which means that, quite apart from the laws of physics and chemistry, as laid down in quantum theory, we must also consider laws of quite a different kind. But even here I do not really know whether we need greater freedom than we already enjoy thanks to the concept of complementarity. As far as I can see, it makes very little difference whether—as in the statistical interpretation of thermodynamics—we join new concepts to quantum mechanics and formulate new laws with them but do not change quantum mechanics itself, or whether, as happened in the extension of classical physics into quantum theory, we extend the theory itself. The real problem is: How can that part of reality which begins with consciousness be combined with those parts that are treated in physics and chemistry? How is it possible that the laws governing these several parts should not conflict? Here we obviously have a genuine case of complementarity, one that we shall have to analyze in greater detail once we know more about biology than we do at present."

And so the conversation continued. For a time Niels manned the wheel and Chievitz took charge of the compass while I continued to look for lights on the black horizon. It was long past midnight, and from time to time a bright patch behind the fairly dense clouds would reveal the position of the moon. We had done a good twenty miles since we had entered Storström Strait, and we ought to have been approaching Omö Sound, which we intended to pass before dropping anchor. According to our map, the entrance to the sound was marked by a besom sticking out of the water, and I could not, for the life of me, imagine how, after sailing some twenty miles in a weak current and in the pitch dark, I was expected to spot a besom.

"Heisenberg, have you found it yet?" Chievitz demanded.

"No, you might just as well ask whether I have found a Ping-Pong ball someone dropped from the last steamer to pass here."

"You are a rotten sailor."

"Why don't you take over then?"

Chievitz raised his voice to make sure he would be heard in the cabin: "It's always the same old story, just like a bad novel: the captain is asleep, the ship hits a reef and the whole crew goes down."

From below we could hear Bjerrum's sleepy voice: "Have you at least some rough idea of where we may be?"

Chievitz answered: "Yes, we can tell precisely; we are on the yacht *Chita*, under Captain Bjerrum, who is fast asleep."

Bjerrum now came up and took over. In the dim distance, I

could see the flashes of a beacon from which we had to take our bearings very carefully. I was told to take soundings with a plumb line, which I managed to do fairly accurately, since we were making rather slow progress. We consulted our map and, from the position of the beacon and the measured depth, were able to establish, to everyone's great relief, that we must be just half a mile from the besom I had been looking for.

We sailed on for a few more minutes. Bjerrum joined me forward, and, while I saw nothing but undiluted darkness, he suddenly called: "There it is." The entrance to Omösund was just a few hundred yards away. We dropped anchor on the far side of the island, and all of us were glad to spend the rest of the night fast asleep in the cabin.

10

Quantum Mechanics and
Kantian Philosophy (1930–1934)

My Leipzig circle grew very quickly. Highly talented young men from many different countries joined us either to take part in the development of quantum mechanics or else to apply the new theory to the structure of matter. And these active, open-minded physicists all helped to enliven our seminar, and almost monthly widened the sphere of application of the new ideas. Felix Bloch from Switzerland did much to increase our understanding of the electric properties of metals; Lev Landau from Russia and Rudolph Peierls examined the mathematical problems of quantum electrodynamics, Friedrich Hund developed the theory of chemical bonds, Edward Teller determined the optical properties of molecules. Carl Friedrich von Weizsäcker joined this group at the age of only eighteen and introduced a philosophical note into our discussions; although he was a student of physics, he grew unusually animated whenever our talks impinged on philosophical or epistemological problems.

We were offered a special occasion for philosophical discussions one or two years later when the young philosopher Grete Hermann came to Leipzig for the express purpose of challenging the philosophical basis of atomic physics. In Göttingen, she was an active member of the circle around the philosopher Leonard Nelson, and thus steeped in the neo-Kantian ideas of the early-nineteenth-century philosopher and naturalist Jakob Friedrich Fries. One of the requirements of Fries' school and hence of Nelson's circle was that all philosophical questions must be treated with the rigor normally reserved for modern mathematics. And it was by following this rigorous approach that Grete Hermann believed she could prove that the causal law—in the form Kant had given it—was unshakable. Now the new quantum mechanics seemed to be challenging the Kantian conception, and she had accordingly decided to fight the matter out with us.

Her first discussion was with Carl Friedrich von Weizsäcker and myself, and probably went something like this:

"In Kant's philosophy," she began, "the causal law is not an empirical assertion which can be proved or disproved by experience, but the very basis of all experience—it is part of the categories of the understanding Kant calls 'a priori.' The sense impressions by which we grasp the world would be nothing but subjective sensations, with no objective correlates, if there were not a rule by which certain impressions must follow from certain preceding ones. This rule, i.e., the existence of a strict relationship between cause and effect, must be presupposed if we wish to objectify our observations, if, indeed, we wish to assert that we have experienced any thing or process. Now science deals precisely with objective experiences; only experiences that can also be verified by others, and are objective in precisely that sense, can be the objects of natural science. It follows that science must presuppose a causal law, that science itself can exist only because there is such a law. The causal law is a mental tool with which we try to incorporate the raw material of our sense impressions into our experience, and only inasmuch as we manage to do so do we grasp the objects of natural science. That being the case, how can quantum mechanics possibly try to relax the causal law and yet hope to remain a branch of science?"

I tried to describe the experiments that had led to the statistical interpretations of quantum theory:

"Let us take a single atom of Radium B. It is, of course, very much easier to experiment with a large number of such atoms, i.e., with a small chunk of Radium B, than with a single atom, but in principle there is no reason why we should not also study the behavior of the latter. We know that, sooner or later, the Radium B atom must emit an electron in some direction and change into a Radium C atom. On the average this will happen after half an hour or so, but a particular atom may become transformed after seconds or only after days. By 'average' we simply mean that, in the case of large numbers of Radium B atoms, half of them will have become transformed after about thirty minutes. But we cannot—and this is where the causal law breaks down—explain why a particular atom will decay at one moment and not at the next, or what causes it to emit an electron in precisely this direction rather than that. And we are convinced, for a variety of reasons, that no such cause exists."

"That is precisely," Grete Hermann said, "where so many people think modern physics has gone wrong. The mere fact that no cause for a certain effect has yet been discovered does not mean that no such cause exists. I myself would simply conclude that the problem has still to be solved, that atomic physicists must go on searching until they discover the cause. After all, your knowledge of the state of the Radium B atom before the emission of the electron is incomplete inasmuch as you cannot tell when and in what direction the electron will be emitted. In other words, you will have to keep looking."

"No, we think that we have found all there is to be found in this field," I insisted, "for from other experiments with Radium B we know that there are no determinants beyond those we have established. Let me put it more precisely: we have just said that it is impossible to tell in which direction an electron will be emitted, and you tell us to keep looking for further factors. Even assuming that you were right and we could discover such factors, we should get into new difficulties. You see, the electron can also be treated as a material wave sent out by the atomic nucleus. Such a wave can cause interference phenomena. Let us further assume that those parts of the wave which the atomic nucleus emits in opposite directions can be made to interfere within a special apparatus. The result will be extinction in certain directions. In that case we could make the certain prediction that the electron will not ultimately be emitted in that direction. But if we had discovered new determinants from which we could tell that the electron was originally emitted in a clearly defined direction, then no interference could have occurred. There would be no extinction, and our earlier conclusion would have been wrong. In fact, however, extinction can be observed experimentally. This is nature's own way of telling us that no new determinants exist, that our knowledge is complete without them."

"But this is dreadful," Grete Hermann said. "On the one hand, you claim that your knowledge of the Radium B atom is incomplete inasmuch as you cannot tell when and in what direction the electron will be emitted, while, on the other hand, you tell me that your knowledge is complete because, if there were further determinants, you would get into trouble with other experiments. But our knowledge cannot possibly be complete and incomplete at the same time. The whole thing is sheer nonsense."

Carl Friedrich now joined the discussion. "The apparent contradiction," he said, "probably arises because we behave as if a Radium B atom were a 'thing-in-itself,' a Kantian *'Ding an sich.'* But this is by no means self-evident or correct. Even Kant treated the *'Ding an sich'* as a problematical concept. He realized that we can say nothing about the *'Ding an sich'* as such; we can only speak about the objects of our perception. He did, however, assume that we could correlate or arrange these objects according to the model of the *'Ding an sich.'* In other words, he treated as a priori that structure of experience to which we have become used in daily life and which, in more precise form, is also the basis of classical physics. In this view the world consists of things in space that change with time, of processes that follow one another according to a set rule. But in atomic physics we have learned that observations can no longer be correlated or arranged on the model of the *'Ding an sich.'* Hence there is also no 'Radium B atom in itself.' "

Grete Hermann cut him short. "You don't seem to be using the term *'Ding an sich'* in the spirit of Kantian philosophy. We must clearly distinguish between the *'Ding an sich'* and the physical object. According to Kant, the *'Ding an sich'* does not appear in phenomena, not even indirectly. This concept has only one function: in natural science and in the whole of theoretical philosophy it refers to what cannot possibly be known. Because, you see, our entire knowledge is based on experience, and experience means knowing precisely things as they appear to us. Even a priori knowledge does not deal with things as they are in themselves; its only function is to make experience possible. Now if you refer to a 'Radium B atom *an sich*,' in the sense of classical physics, you are simply referring to what Kant calls a thing or an object. Objects are part of the world of phenomena: chairs and tables, stars and atoms."

"Even if we cannot see them, like atoms, for instance?"

"Even then, for we deduce them from observable phenomena. The world of phenomena is a coherent structure, and it is quite impossible, even in daily experience, to distinguish sharply between what we see directly and what we merely infer. You see this chair, but you can't see its back from where you are standing. Still, you are positive that it exists. This simply means that science is objective; it is objective precisely because it speaks of objects, not of perceptions."

"But when it comes to atoms, we can see neither front nor back. Why should they have the same properties as chairs and tables?"

"Because they *are* objects. Without objects there can be no objective science. And what objects are is determined by such categories as substance, causality, etc. If you renounce the strict application of these categories, then you also renounce the possibility of experience in general."

But Carl Friedrich refused to give in. "In quantum theory we have to use a new method of objectifying perceptions, one that Kant would never have dreamt of in his philosophy. Every perception refers to an observational situation that must be specified if experience is to result. The consequence of a perception can no longer be objectified in the manner of classical physics. Once an experiment allows us to deduce the presence of a Radium B atom, then the resulting knowledge is complete for this particular observational situation, but incomplete for another observational situation, for example, one involving statements about the emission of an electron. If two observational situations are in the relationship Bohr has called complementary, then complete knowledge of one necessarily means incomplete knowledge of the other."

"And with that you intend to overthrow the whole Kantian analysis of experience?"

"No, I do not. Kant has perceived very shrewdly how we come

by our experiences, and I believe that his analysis is essentially correct. But when he makes the intuitive forms 'space' and 'time,' and the category 'causality,' a priori conditions of experience, he runs the danger of postulating them as absolutes and of claiming that they must enter into the content of all physical theories. But this is not the case, as relativity and quantum theory have shown. Nevertheless Kant is perfectly right in his own way. Physical experiments must first of all be described in the language of classical physics, or else the results cannot be communicated to other physicists who have to verify them. The Kantian 'a priori' is therefore by no means eliminated from modern physics; it has simply been 'relativized.' The concepts of classical physics, which include space, time and causality, may be said to be a priori conditions of relativity and quantum theory, inasmuch as they must be used in the description of experiments—or let us put it more circumspectly, inasmuch as they are actually used in that way. But their content is nevertheless changed by the new theories.''

"All this still fails to provide an answer to my original question," Grete Hermann said. "I was wondering why our inability to discover the causes of, say, the emission of an electron means that we must stop searching further. Admittedly, you don't forbid this search, but you claim that it is futile since no further determining factors can be found. Indeed, you contend that, if only it is formulated in precise mathematical language, the indeterminacy allows a definite prediction in another experiment. And this, too, you claim is borne out by the results. If you argue like that, you turn uncertainty into a physical reality, with an objective character, while, normally speaking, uncertainty is a synonym for ignorance, and as such something purely subjective."

Here I felt I must once again intervene in the discussion. "With your last remark," I said, "you have given a very precise description of the most characteristic feature of modern quantum theory. Whenever we try to deduce laws from our study of atomic phenomena, we discover that we no longer correlate objective processes in space and time, but only observational situations. Only for these can we derive empirical laws. The mathematical symbols with which we describe such observational situations represent possibilities rather than facts. One might say that they represent an intermediate stage between the possible and the factual, which can only be called objective in the sense that, say, temperature is called objective by statistical thermodynamics. Our knowledge about what is possible does admittedly enable us to make a few clear predictions, but, as a rule, it only allows us to speculate as to the probability of a future event. Kant could not possibly have foreseen that in an experimental realm so far beyond daily experience we could no longer treat observations as if they referred to *'Dinge an sich'* or 'objects'; in other words, he could not foresee that atoms are neither things nor objects."

"In that case what are they?"

"We lack the right term, for our language is based on daily experience, and atoms are not. But if you dislike such evasions, we might say that atoms are parts of observational situations, parts that have a high explanatory value in the physical analysis of the phenomena involved."

"Since we are talking about linguistic difficulties," Carl Friedrich now interjected, "it is worth remembering that the most important lesson we can learn from modern physics is perhaps the fact that all the terms with which we describe experience apply to a limited realm only. All such concepts as 'thing,' 'object of perception,' 'moment,' 'simultaneity,' 'extension,' etc., get us into trouble in certain experimental situations. That does not mean that these concepts have ceased to be the presupposition of all experience, but it does mean that they are the presupposition that must be critically evaluated in each case, and from which no absolute rules can be deduced."

Grete Hermann seemed very unhappy about this turn in our conversation. She had set out to refute the arguments of atomic physics with Kantian propositions or, conversely, had hoped to be shown that Kant had been guilty of a serious philosophical lapse. But what she now found was a ramshackle halfway house, in which she did not really feel at home. Hence she asked: "Is your relativization of the Kantian 'a priori,' indeed of language itself, not tantamount to complete resignation in the sense of 'I see that nothing can be known'? Do you finally believe that there is no ground of knowledge on which we can safely take our stand?"

Carl Friedrich answered very boldly that it was precisely the development of natural science that seemed to justify a slightly more optimistic view:

"When we say that, with his 'a priori,' Kant gave a correct account of the state of scientific knowledge in his time, but that modern atomic physics faces a new epistemological situation, then our statement may be compared to the other statement, that Archimedes' lever laws were the right formulation of the practical rules of technology in Archimedes' day, but do not meet the needs of modern technology, for instance of electronics. Archimedes' laws represent true knowledge, not some vague expression of opinion. They apply to all levers at all times, and if there should be life on some planet of some distant stellar system, then Archimedes' laws will apply to them as well. The fact that extensions of knowledge have helped us to advance into realms of technology in which the lever concept no longer suffices signifies neither the relativization nor the historization of the lever laws; it simply means that in the course of historical development these laws have lost the central significance they originally enjoyed. Similarly, I believe that Kant's analysis of human understanding represents true knowledge, not some vague expression of opinion, and that it will apply whenever thinking beings enter into the kind of contact with their environment to which we refer as 'experience.' But even the Kantian 'a priori' can be displaced from its central position and become part of a very much wider analysis of the process of understanding. In this context, it would certainly be a mistake were we to detract from scientific or philosophical knowledge with the phrase, 'Every age has its own truth.' We should nevertheless remember that the very structure of human thought changes in the course of historical development. Science progresses not only because it helps to explain newly discovered facts, but also because it teaches us over and over again what the word 'understanding' may mean."

This reply, based partly on Bohr's teachings, seemed to satisfy Grete Hermann to some extent, and we had the feeling that we had all learned a good deal about the relationship between Kant's philosophy and modern science.

11

Discussions about Language (1933)

The golden age of atomic physics was now fast drawing to an end. In Germany political unrest was increasing. Radical groups of the right and the left came out into the streets, fought in the backyards of the poorer quarters and tried to break up each other's meetings. Almost imperceptibly, tension mounted, even at the university and during faculty meetings. For a time I tried to close my eyes to the danger, to ignore the ugly scenes in the street. But, when all is said and done, reality is stronger than all our wishes—this time it entered my consciousness in the form of a dream. One Sunday morning I was due to go on a bicycle tour with Carl Friedrich, and I had set the alarm clock for five o'clock. Just before I woke up, I saw a strange vision in my half sleep. I was walking up the Ludwigstrasse in Munich at first light, just as I had done in the spring of 1919. The street was bathed in a reddish, increasingly intense and uncanny glow. Crowds of people with scarlet and black-red-and-white flags were streaming from the Victory Gate toward the university fountains and the air was filled with noise and uproar. Suddenly, just in front of me, a machine gun began to cough. I tried to jump to safety and woke up; the sputtering of the gun was simply the ringing of my alarm clock, and the reddish light the morning sun on my bedroom curtains. From that moment I knew that we were once again facing hard times.

After the catastrophe of January 1933 I took just one more happy holiday with old friends, long remembered by all of us as a beautiful but painful farewell to the "golden age."

I had the use of a skiing hut on an alpine meadow high above the village of Bayrischzell on the southern slope of the Grosse

Traithen. It had been restored by friends from the Youth Movement after an avalanche had half-destroyed it. The father of one of my comrades, a timber merchant, had supplied the necessary materials and tools, the farmer who owned the hut had carried them up to the meadow during the summer, and within a few glorious autumn weeks my friends had put up a new roof, repaired the shutters and fixed up a dormitory inside. As a reward, all of us were allowed to use the hut as a skiing hostel, and for Easter 1933 I invited Niels and his son Christian, Felix Bloch and Carl Friedrich for a skiing holiday. Niels, Christian and Felix decided to come to Oberaudorf straight from Salzburg, where Niels had an engagement, and climb the rest of the way. Carl Friedrich and I had gone up two days earlier to fix things up and to lay in provisions. A few weeks earlier, during the good weather, cases of food had been delivered to the Brünnstein refuge, whence we had to carry them up to our hut, just under an hour's walk away, in rucksacks. During our first night in the mountains a gale blew up and there was a continuous fall of snow. In the morning, we had great difficulty in clearing the entrance, and by noon when we fought our way through the blizzard to the refuge through new snow almost a yard deep, the weather showed no sign of breaking, and we realized that we had to watch out for possible avalanches. From the refuge I rang up Niels in Salzburg, as we had agreed, described conditions on the mountain and promised that Carl Friedrich and I would meet him at Oberaudorf station next day. Niels thought that this was quite unnecessary; he, Christian and Felix would simply take a taxi from Oberaudorf and come up to the hut. I had to explain that this idea was extremely unrealistic, and then he agreed to my original suggestion. During the second night, too, it kept snowing hard, and by morning our hut was almost buried. Yesterday's tracks had completely disappeared. Luckily, the sky began to clear, visibility was good, and we were able to avoid the worst spots. Carl Friedrich and I took turns at digging a new path to the refuge; once there we could ski downhill to Oberaudorf all the way. Our tracks would help us to retrace our path—the sky was clear and the wind still, so that there was no danger of a snowdrift before nightfall. But when we reached the station at the agreed time, there was no sign of Niels, Christian or Felix. Instead, a great deal of luggage was being unloaded: skis, rucksacks, coats, all of which looked very much as if they belonged to our guests. We learned from the stationmaster that the owners of the luggage had missed their train because they had insisted on taking coffee at an intermediate stop and that they would not be arriving before 4 P.M. at the earliest. This meant that most of our difficult climb would have to be made in the dark. Carl Friedrich and I used the interval to sort out the essential luggage—we had to conserve our physical strength on the way up. When our guests arrived punctually at four o'clock, I told Niels that we were in for quite an adventure—the tracks Carl Friedrich and I had left coming downhill were our only signpost through the thick snow.

"Strange, isn't it," Niels said after a moment's reflection, "and there I was thinking that a mountain is something you have to start climbing from the bottom."

One of us mentioned that something like "inverse mountaineering" was the rule in the Grand Canyon. You got there by train, were put down by the edge of a vast desert plateau, had to descend six thousand feet to the Colorado River and then climb back to the train. That was precisely why the place was called a canyon and not a mountain. Chatting away like this, we made good progress during the first two hours, but I kept remembering that a climb that takes no more than two to three hours in summer might well take six or even seven under the present conditions. It was pitch-dark when we came to the most difficult part of our ascent. I went ahead, followed by Niels, Carl Friedrich who was carrying a lantern, Christian and Felix, in that order. Most of our tracks were still deeply chiseled into the snow and hence easy to find—only in a few unprotected spots had the wind covered them up. I was rather uneasy when I found that the snow had remained so powdery; Niels was beginning to tire and we had to slow down. It was ten o'clock now, and I thought it would take us another hour to reach the refuge.

As we passed a steep slope, something very odd happened—I suddenly had the feeling that I was swimming. I completely lost control of my movements, and then something pressed on me so violently from all sides that, for a moment, I stopped breathing. Luckily, my head had stayed above the encroaching masses of snow, and within seconds I got my arms free again. I turned around. It was completely dark and none of my friends was in sight. I called out "Niels" but received no answer. For a moment, I thought all of them had been buried in the avalanche. I made frantic exertions to dig my skis out of the snow, and when I had done so, I spotted a light a long way up. I yelled for all I was worth and received an answer from Carl Friedrich. It suddenly dawned on me that I had been carried quite a long way down by the avalanche, and that the others had been spared. I now made my way toward the lantern, and we continued with the utmost caution. At eleven, we reached the refuge and decided not to run the risk of a further climb that night. We accordingly bedded down, and early next morning reached our hut after laboring through great drifts of blinding white snow under a dark-blue sky.

Too exhausted by the climb and the shock of the avalanche, we took it easy during that day. We lay on the roof, from which we had cleared the snow, enjoying the sun and discussing recent developments in atomic physics. Niels had brought along a cloud-chamber photograph from California, which immediately captured our interest and gave rise to heated discussions. The argument hinged on a problem Paul Dirac had brought up a few years earlier in connection with his work on relativistic electron theory. According to that theory, which had meanwhile been borne out by experiment, there were mathematical reasons for concluding that, in addition to the negatively charged electron, there must also be a related particle with a positive charge. Dirac had at first tried to identify this hypothetical particle with the proton, that is, with the atomic nucleus of the hydrogen atom, but most other physicists had objected to this hypothesis on the ground that there was convincing evidence to show that this positively charged particle ought to have the same mass as the electron, while the proton was known to be some two thousand times heavier. In addition, the hypothetical particle was said to behave quite differently from usual matter—upon collision with an ordinary electron the two were supposed to be transformed into radiation. (Today we speak, accordingly, of "antimatter.")

Now Niels' cloud-chamber photograph seemed to prove the existence of just such an (anti) particle. It showed a trail of water droplets obviously produced by a particle coming from the top. The particle had then crossed a lead plate and had left a further trail on the other side of the plate. The cloud chamber had been placed in a strong magnetic field so that the tracks had become deflected. The density of the water droplets in the track corresponded precisely to the density one could expect from electrons, but the deflection suggested the presence of a positive electrical charge, that is, if the particle had indeed come from the top. Now this followed necessarily from the fact that the curvature above the plate was smaller than that below, in other words, from the fact that the particle had been slowed down by the lead plate. We now argued at length whether or not these conclusions were necessarily correct. All of us realized that the answer was crucial.

After we had been looking for possible experimental mistakes for some time, I said to Niels: "Isn't it odd that, throughout this discussion, no one should have mentioned quantum theory? We behave as if the electrically charged particles were an object like an electrically charged oil droplet, or like a pith ball in an old electroscope. We quite unthinkingly use the concepts of classical physics, as if we had never heard of the limitations of these concepts and of uncertainty relations. Isn't that bound to lead to errors?"

"No, certainly not," Niels replied. "After all, it is almost the essence of an experiment that the observations can be described with the concepts of classical physics. That is the whole paradox of quantum theory. On the one hand, we establish laws that differ from those of classical physics; on the other, we apply the concepts of classical physics quite unreservedly whenever we make observations, or take measurements or photographs. And we have to do just that because, when all is said and done, we are forced to use language if we are to communicate our results to other people. A measuring instrument is a measuring instrument only when the observations it yields enable us to arrive at unequivocal conclusions about the phenomenon under observation, only when a strict causal connection can be assumed to exist. Yet when it comes to the theoretical description of an atomic phenomenon, we must make a distinction between the phenomenon and the observer or his apparatus. The demarca-

tion line may be subject to choice, but on the observer's side of
the split we are forced to use the language of classical physics,
simply because we have no other language in which to express
the results. We know that the concepts of this language are
imprecise, that they have a limited area of application, but we
have no other language, and, after all, it does help us to grasp
the phenomenon at least indirectly."

"Isn't it possible," Felix objected, "that once we have understood quantum theory even better, we might be able to dispense
with the classical concepts and use a new language to speak far
more accurately about atomic phenomena than we can today?"

"You are misunderstanding the problem," Niels replied. "Science is the observation of phenomena and the communication of
the results to others, who must check them. Only when we have
agreed on what has happened objectively, or on what happens
regularly, do we have a basis for understanding. And this whole
process of observation and communication proceeds by means of
the concepts of classical physics. The cloud chamber is a measuring apparatus, which means that this photograph entitles us to
conclude that a positively charged particle which has the properties of an electron has passed through the chamber. Of course,
we have to assume that the measuring instruments were properly
built, that they were firmly fixed to the table, that the camera
was mounted so rigidly that it did not shake when the photograph was taken, that the lens was properly focused, etc.; in other
words, we must be certain that the experiment was performed
under the precise conditions laid down by classical physics. It is
one of the basic presuppositions of science that we speak of
measurements in a language that has basically the same structure
as the one in which we speak of everyday experience. We have
learned that this language is an inadequate means of communication and orientation, but it is nevertheless the presupposition
of all science."

While the rest of us continued sunning ourselves on the roof
and engaging in physical and philosophical discussions, Christian
decided to explore the immediate vicinity of our hut. He
brought back a damaged wind wheel which some of my Youth
Movement friends must have built during an earlier stay—perhaps to determine the strength and direction of the wind, or
perhaps purely for fun.

We at once decided to construct a new and better wheel. Niels, Felix and I each began to carve one from bits of firewood. But while Felix and I tried to produce the perfect hydrodynamic shape—that is, a propeller—Niels took a square piece of timber and carved out two planes set at right angles. We discovered that our "ideal" propeller was mechanically so unsound that it would not revolve properly in the wind, while Niels' simple instrument was built so well, down to its shaft, that it turned in the slightest breeze.

"You gentlemen are far too ambitious" was all Niels said of our efforts, though he himself had, of course, been no less ambitious in his clean and careful craftsmanship, which, incidentally, reflected his whole attitude to classical physics.

That night we played poker. We could, of course, have listened to the bad phonograph, and to the awful music-hit records stored in the hut, but the demand for this type of music was very slight. Our poker game was of a rather original variety. The hands on which we staked our money were shouted out and praised to the skies in an attempt to outbluff each other. Niels saw this as a fresh opportunity for philosophizing about the meaning of language.

"It is quite obvious," he said, "that in this game we are using language quite differently than we do in science. To begin with, we try to hide rather than bring out the real facts. Bluffing is part of the game. But how do we hide the real facts? Language may convey pictures to others that help to oust ideas reached by sober reflection and so give rise to mistaken actions. But what factors decide whether or not these pictures impinge on others with sufficient intensity? Surely not the loudness of our voices. This would be much too primitive a view. Nor is it the kind of routine persuasion a good salesman might use. For none of us are familiar with this routine, and we can hardly imagine that any of us would be taken in by it either. Perhaps our ability to convince others depends on the intensity with which we can persuade ourselves of the force of our own imagination."

This view was given unexpected confirmation during the game. Niels was loudly insisting that he was holding five cards of the same suit. He kept doubling, and by the time four cards had been turned up the rest of us threw in our hands. Niels won a large sum of play money. When it was all over, and he proudly showed us his fifth card, it turned out that it was not of the same suit, as he himself had wrongly thought. He had mistaken the ten of hearts for a ten of diamonds. Hence his bidding had been pure bluff. I was reminded of his remark, during our walk through Zealand, about the power of ideas to mold men's thoughts through the ages.

Our hut had turned terribly cold, and not even the stiff grog with which we had enlivened our poker game helped us to ignore that fact. And so we all climbed into our sleeping bags and bedded down on our bales of straw. In the quiet of the night, my thoughts began to harken back to Niels' cloud-chamber photograph. Could it be true that the positive electrons predicted by Dirac really did exist, and, if so, what were the consequences? The more I thought about it, the more strongly I was gripped by the kind of emotion that always seizes one when one is forced to change one's opinions on fundamental points. During the past year I had been working on the structure of the atomic nucleus. Chadwick's discovery of the neutron had suggested that atomic nuclei consist of protons and neutrons, held together by strong but hitherto undiscovered forces. That seemed plausible enough. The further assumption that the nucleus was completely devoid of additional electrons had seemed very much more questionable, and some of my friends had criticized me most sharply for propounding this view. "After all," they had said, "you can see electrons leaving the atomic nucleus during beta disintegrations." But I had imagined that the neutron itself was made up of a proton and an electron, even though, for reasons I did not yet understand, it was no larger than the proton. Moreover, the newly discovered forces responsible for holding the atomic nucleus together did not seem to change when a proton was replaced with a neutron. This symmetrical interaction could be explained partly by the assumption that the cohesive force resulted from the exchange of an electron between the two heavy particles. But this picture had two major flaws. To begin with, there was no real reason why there should not be equally strong forces binding proton to proton, or neutron to neutron. And then it was inexplicable why—apart from the relatively small electrical contribution—the two forces should appear to be identical. Moreover, the neutron was empirically so much

like the proton that it seemed unreasonable to consider the one
as a simple and the other as a compound structure.

But if Dirac's positive electron—or, as we now call it, the
positron—really existed, then we were faced with a completely
new situation: the proton could now be imagined to consist of a
neutron and a positron, and the symmetry between proton and
neutron was completely restored. But, in that case, was there any
sense in asserting that electrons or positrons were contained in
the atomic nucleus? Could they not have been produced from
energy, as a converse to Dirac's assertion that they combined to
produce radiation? And if energy could be transformed into
electron and positron pairs and vice versa, could we still ask how
many particles went into the atomic nucleus? So far we had
always believed in the doctrine of Democritus, which can be
summarized by: "In the beginning was the particle." We had assumed that visible matter was composed of smaller units, and
that, if only we divided these long enough, we should arrive at
the smallest units, which Democritus had called "atoms" and
which modern physicists called "elementary particles." But perhaps this entire approach had been mistaken. Perhaps there was
no such thing as an indivisible particle. Perhaps matter could be
divided ever further, until finally it was no longer a real division
of a particle but a change of energy into matter, and the parts
were no longer smaller than the whole from which they had been
separated. But what was there in the beginning? A physical law,
mathematics, symmetry? In the beginning was symmetry! This
sounded like Plato's *Timaeus,* and I was reminded of the day I
spent on the roof of the theological college in the summer of
1919. If the particle in the cloud-chamber photograph was really
Dirac's positron, vast new realms had been opened up, and I
could already discern some of the paths on which we should have
to advance into them. And so I kept musing until, at long last, I
fell asleep.

Next morning the sky was blue again. We put on our skis right
after breakfast, and went up the Himmelmoos slope to a small
lake and on across a ridge into a lonely valley behind the Grosse
Traithen and from there back to the top of our mountain. On
the crest running eastward we suddenly witnessed a strange
meteorological and optical phenomenon. The gentle wind blow-

ing from the north carried up a thin cloud of vapor brightly illuminated by the sun; in it we could see our shadows, and those cast by our heads were surrounded by a bright ring of light. Niels, who seemed particularly delighted by this unusual spectacle, said that he had heard about it before. People had told him that it was possibly the origin of the halo in which the early masters had wreathed the heads of the saints. "And perhaps it is characteristic," he added with a wink, "that it is only around our own heads that we can see the halo." This remark was greeted with great glee and also gave rise to quite a few self-critical remarks. Then it was time for a quick race down to our hut. Felix and I were particularly ambitious, and so wildly did I set off that, cutting into a steep slope, I started another avalanche. Fortunately, we managed to stay on top of it, and all of us arrived home safely, although at considerable intervals. It was now my job to cook the meal, and Niels, who was rather tired, sat down with me in the kitchen, while the rest sunned themselves on the roof. I used the opportunity to continue our earlier conversation.

"Your explanation of the halo," I said, "is most convincing, and I am prepared to accept it as at least part of the truth. But I am only half-satisfied, and once, in a letter to an overzealous positivist of the Vienna Circle, I put forward the opposite viewpoint. The positivist assertion that every word has a clear meaning and that it is quite improper to use it in any other way struck me as arrant nonsense. Accordingly, I wrote that everyone knows what is meant when we say that a great man sheds light wherever he goes. I realized, of course, that this sort of light could not be measured with a photometer, but I refused to take the physical meaning of the word 'light' as the real one, and dismiss the other as purely derived. Quite possibly, therefore, it was the light of the saints themselves and not the strange phenomenon we have just observed that inspired the painters."

"I am quite prepared to accept that," Niels told me, "and we agree much more than you think. Of course, language has this strange, fluid character. We never know what a word means exactly, and the meaning of our words depends on the way we join them together into a sentence, on the circumstances under which we formulate them, and on countless subsidiary factors. If you read the American philosopher William James, you will find that he has described it all most accurately. He says that, though our minds may seem to seize on only the most important meaning of a word we hear spoken, other meanings arise in its darker recesses, link up with different concepts and spread into the unconscious. That happens with everyday speech and a fortiori with the language of the poets. To a lesser extent, it applies to the language of science as well. Particularly in atomic physics, nature has taught us that some of our most trusted concepts have a strictly limited application. You have only to think of position and velocity.

"In spite of all that, Aristotle and the ancient Greeks took a great step forward when they discovered that language can be idealized and rendered precise enough for logical deductions. That kind of language is, of course, much narrower than everyday speech, but it is of inestimable value in natural science.

"The positivists are quite right when they stress the importance of linguistic accuracy and when they warn us that language may become meaningless once it eschews logical rigor. But perhaps they overlook the fact that in science we can at best try to approximate this ideal, but can never actually attain it. For the language with which we describe our experiments contains concepts whose scope we cannot define with precision. One could, of course, say that the mathematical formulae with which we theoretical physicists describe nature ought to have this degree of logical purity and strictness. But then the whole problem reappears in different guise just as soon as we try to apply these formulae to nature. For if we want to say anything at all about nature—and what else does science try to do?—we must somehow pass from mathematical to everyday language."

"The positivists," I remarked, "seem to aim their shafts chiefly at metaphysics and especially at religious metaphysics. To them, the arguments of religion are nothing but pseudo problems which cannot stand up to linguistic analysis, and are therefore meaningless. Do you think they are right, if only in part?"

"Certainly their critique contains a large grain of truth," Niels replied, "and we can learn a great deal from it. I do not object to positivism on the grounds that I would be less skeptical in this area, but rather because I am afraid that, on principle, things cannot be much better in science either. To put it in an exaggerated way: in religion we renounce the wish to give words an unequivocal meaning from the outset, while in science we start with the hope—or, if you like, the illusion—that one day it may be possible to do just that. But for all that, we can learn a great deal from the positivists. For instance, I cannot see what people have in mind when they speak about the meaning of life. After all, the word 'meaning' refers to a connection between a subject or object and something else, for instance, an intention, an idea or a plan. But when it comes to life—by that I mean the whole of life, the world we experience—to what else can we possibly refer it?"

"But, surely, we do know what we mean when we speak of the 'meaning' of life," I objected. "The meaning of life depends on ourselves. I think the expression refers to the way we shape our own life, in which we fit it into a wider context; perhaps it is only an image, a principle, a faith, but still something we can fully understand."

Niels reflected in silence a while, and then said: "No, the meaning of life is simply that there is no meaning in saying that life has no meaning. Our quest for understanding is like a well without a bottom."

"But aren't you being too rigid with language? You know that the old Chinese sages placed the 'tao' at the head of all philosophy, and 'tao' is often translated as 'meaning.' The Chinese sages would probably have raised no objection to linking the words 'tao' and 'life.'"

"If you give the term 'meaning' so wide a definition, then anything is possible. But none of us can really say how the word 'tao' was used. Still, since you are talking about Chinese philosophers and life, I must say I prefer the old legend about the three sages who were asked to describe the taste of vinegar. Perhaps I ought to add that the Chinese call vinegar the 'water of life.' The first philosopher said, 'It is sour'; the second, 'It is bitter'; and the third, none other than Lao-tse, 'It is fresh.'"

Carl Friedrich came into the kitchen and asked when in God's name I was going to serve up the food. I told him that if he called the others, and put out the aluminum plates and cutlery, we could start eating at once. We sat around the table and the old proverb "Hunger is the best cook" proved my salvation. After the meal, we established a roster of duties: Niels would wash up, I would clean the stove, the others would chop wood or sweep the hut. It goes without saying that our primitive kitchen would have caused a sanitary inspector's hair to stand on end. "Our washing up is just like our language," Niels said. "We have dirty water and dirty dishcloths, and yet we manage to get the plates and glasses clean. In language, too, we have to work with unclear concepts and a form of logic whose scope is restricted in an unknown way, and yet we use it to bring some clarity into our understanding of nature."

During the next few days the weather proved changeable and we went on a number of excursions, some long, others short. We climbed the Trainsjoch and did skiing exercises on the training slope of the Unterberger-Alm. Once again our talk came round to the problem of language, this time after Carl Friedrich and I had tried to take snapshots of a herd of chamois one afternoon. We failed to get close enough, and admired the instinct of animals who could detect the softest sound, the merest footprint in the snow, the crackling of a branch or the slightest scent, and take evasive action. This gave Niels cause for a disquisition on the difference between intellect and instinct.

"Perhaps these chamois only succeeded in escaping from you because they did not have to think first, or discuss the best method of eluding you. Their whole organism is specialized for finding safety on mountainous terrain. Selection no doubt helps a particular species to develop certain physical capacities to near-perfection. As a result, however, it is forced to wage the struggle for life in a particular manner. If environmental conditions change too much, it can no longer adapt itself and will become extinct. There are some fish that can produce electric shocks and so defend themselves against enemies. There are others whose appearance is so perfectly adapted to life in the sand that they completely merge with the sea bed and so fool all predators. With us human beings specialization has taken a different path. Our nervous system, which enables us to think and speak, can be considered an organ with which we can probe much deeper into time and space than any other animal. We can remember the past and predict probable future events. We can also imagine what happens far away in space, and we can make use of the experiences of others. As a result, we have become much more flexible, much more adaptable than any other animal, to such an extent that one can speak of man's specialization toward flexibility. But quite naturally this development of thought and speech, or, more generally, the preponderance of the intellect, has led to a stunting of our instinctual reactions. In that respect, therefore, man is inferior to the animals. He does not have as keen a sense of smell and he cannot jump across rocks as sure-footedly as a chamois. But he can compensate for these deficiencies by reaching into wider spatial and temporal spheres. Here, the development of language was probably the decisive step. For speech, and with it thought, is an ability which—in contrast to all other physical capacities—does not develop within individuals but between individuals. We learn our speech from others. Language is, as it were, a net spread out between people, a net in which our thoughts and knowledge are inextricably enmeshed."

"If you hear a positivist or a logician speak about language," I broke in, "you get the impression that the forms and expressive power of language can be treated and analyzed quite regardless of evolution and biological precedent. Yet if one compares intellect and instinct, as you have just done, it is possible to imagine that different forms of intellect and language could have appeared in different parts of the world. And, in fact, the grammars of different languages are quite distinct, and perhaps differences in grammar may produce differences in logic."

"Naturally, it is possible to have different forms of speech and thought," Niels replied, "just as there are different races or different parts of an organism. But much as all living organisms are constructed in accordance with the same laws of nature, and largely from approximately the same chemical compounds, so the various possibilities of logic are probably based on fundamental forms that are neither man-made nor even dependent on man. These forms must play a decisive part in the selective development of language; they cannot be its mere consequences."

"Let's come back to the difference between the chamois and

ourselves," Carl Friedrich suggested. "It seemed to me that you were arguing that intellect and instinct are mutually exclusive. Do you simply mean that the natural-selection process favors the development of one at the expense of the other or are you thinking rather of a complementary relationship in which the development of one is incompatible with the development of the other?"

"I merely believe that these two ways of finding one's way in the world are radically different. But, needless to say, many of man's actions, too, are determined by instinct. Quite likely, when we judge others by their appearance or expression, or make up our mind about their intelligence or conversational ability, we operate by instinct as well as by experience."

During this conversation, some of us had begun to tidy the rest of the hut—our holidays were coming to an end within a few days. Niels had started shaving. He had been living like a Norwegian lumberman, stuck away in some remote forest, and was astonished to see from the mirror how the razor was changing him back into a professor of physics within a matter of minutes. "I wonder if a cat, also, would look more intelligent after a shave," he now mused.

That evening we played poker again, and since speech, or rather bluffing, played so large a part in our version of the game, Niels suggested that we try the whole thing without cards. Felix and Christian would probably win, he added, because none of us others could match their persuasive powers. The attempt was made, but did not lead to a successful game, whereupon Niels said: "My suggestion was probably based on an overestimate of the importance of language; language is forced to rely on some link with reality. In real poker one plays with real cards. In that case, we can use language to 'improve' the real hand with as much optimism and conviction as we can summon up. But if we start with no reality at all, then it becomes impossible to make credible suggestions."

When our holidays were over, we took the short, western route down into the valley between Bayrischzell and Landl. It was a warm sunny day, and below, where the snow line stopped, liverwort stood in flower between the trees and the meadows were full of yellow primroses. Since our luggage was heavy, we asked the innkeeper to let us have two horses and an old cart. We tried to forget that we were returning to a world full of political troubles. The sky was as bright as the faces of our two young companions, Carl Friedrich and Christian, when we climbed into the cart and drove out into the Bavarian spring.

12

Revolution and
University Life (1933)

When I returned to my Leipzig Institute at the beginning of the
summer term of 1933, the rot had begun to spread. Several of my
most capable colleagues had left Germany, others were preparing
to flee. Even my brilliant assistant, Felix Bloch, had decided to
emigrate, and I myself began to wonder whether there was any
sense in staying on. From this time of painful reflection I remem
ber two conversations particularly well: one with a young National Socialist student, the other with Max Planck.

At the time I was living in a small attic on the top floor of the
Institute. After moving in, I had proudly acquired a grand
piano from Blüthner's of Leipzig, and I would play on it most
evenings, either by myself or with friends who liked chamber
music. Besides that I was taking lessons from the pianist Hans
Beltz, at the College of Music, and so I would sometimes have to
practice at noon as well.

One afternoon I had just left my apartment after an hour's
practice of the Schumann A Minor Concerto and was on my way
to the Institute, when I spotted a young man in the window seat
of the corridor. I had previously noticed him at my lectures,
dressed in brown uniform. He now rose with some embarrass-
ment, and I asked him if he wished to speak to me.

No, he said rather shyly, he had merely been listening to my
music. But since I had asked him, he would be grateful if he
could ask me a few questions. I showed him into my living room,
where he began to pour out his heart.

"I come to your lectures and I know that there is much we can learn from you. But otherwise we have no contact with you at all. I have listened more than once to your playing. I don't very often get a chance to hear music. I also know that you used to be a member of the Youth Movement, like myself. But you never come to youth meetings nowadays, whether they are run by the National Socialist Student Association, the Hitler Youth or anyone else. I myself am a Hitler Youth leader, and I would very much like to see you at one of our meetings. But you act as if you were one of those conservative old professors who live completely in the past and to whom the new Germany is alien, if not anathema. I simply cannot imagine that someone as young and as musical as you are should show such lack of sympathy toward youth, toward those of us who are anxious to do what we can to build a better Germany. After all, we need people with more experience, people who are prepared to help us in the great work of reconstruction. Perhaps you object to the many ugly things that are happening, to the persecution and expulsion of so many innocent people. But please believe me, I myself find these outrages just as repulsive as you do, and I am certain that none of my friends would take part in that sort of thing. Quite likely it is impossible to stop people from going too far in their first flush of excitement after a great revolution, to stop time-servers from climbing onto a successful bandwagon. We can only hope that after a brief spell they will all be thrown out. That is precisely why we need the cooperation of all who want to build more wisely, who, for instance, can help infuse our movement with many of the ideals that used to inspire the Youth Movement. Won't you please tell me why you are keeping your distance from us?"

"If it were merely a question of joining in with young students, working for what I believe is right, I might well offer my support, in both speech and action. But now that such vast numbers of people have been thrown into the arena, the opinion of a few students and professors hardly matters. Moreover, the leaders of this revolution have made quite sure that the nation will turn a deaf ear to reason: they keep pouring scorn on so-called intellectuals, on all those who show greater spiritual discernment than your new masters. And can you really be sure that you are on the right road toward a better Germany? I won't, of course, deny that you personally may have every intention of getting there; but on the whole all we can say with certainty is that the old Germany is being destroyed, and that injustices flourish all around us—everything else is nothing but wishful thinking. If you would simply try to remedy what grievances there are, I would be with you all the way. But what is happening today is something quite different. You must realize that I cannot help you when Germany is being ruined; it's as simple as that."

"Now you are really being unfair. You can't honestly believe that Germany could still be saved with minor reforms. Ever since 1918 things have been going from bad to worse. True enough, we lost the war; our former enemies were stronger, and we ought to have learned our lesson. But what happened? Nightclubs and cabarets sprang up everywhere, and all who showed concern, all who had made efforts or sacrifices, were mocked and derided. 'Why waste your time,' they said, 'when you can enjoy yourself? The war is lost, but there is always alcohol and beautiful women.' And in business life, corruption was rife as never before. When the government ran short of cash, because it had to pay reparations or because the common people had grown too poor to pay their taxes, it simply printed more money. Why not? The fact that many old and weak people were swindled out of their last few pennies and had to starve seemed to worry nobody. The government had enough money; the rich grew richer, the poor poorer. And you have to admit that Jews were involved in some of the worst scandals of recent years."

"And that entitles you to look upon Jews as some special kind of human beings, to treat them shamelessly and to drive a large number of outstanding people from Germany? Why don't you leave it to the courts to punish the guilty, irrespective of religion or race?"

"Simply because it doesn't happen. Justice has long since been turned into political justice, a means of perpetuating the rotten status quo, of protecting the ruling class and letting the rest go to hell. Just look at the mild sentences they used to give some of the worst swindlers. The stench of decay rose from many other places as well. In art exhibitions the most incredible rubbish, the most unwholesome confusion was offered to the public as great art, and if the man in the street refused to marvel, he was told, 'You simply don't understand it; you are too ignorant.' And did the State ever bother about the poor? They boasted about all those wonderful social institutions that were there to ensure no one went hungry. But is it really enough to give the poor just enough money to keep them alive and then forget about them? You must admit, we are doing things very much better. We sit down with the workers, we drill with them in the same Storm Trooper squads, we collect food and clothing for the poor, we march side by side with them in demonstrations, and we can feel that they welcome us. Surely, that's an improvement, or isn't it? During the past fourteen years, everyone worked to line his own pockets. The only thing that mattered was being better dressed than your neighbor, having more elegant quarters, greater pretensions. And Reichstag deputies sought only to gain the maximum material advantage for their own party. Everyone called everyone else greedy, only to become greedier himself. Meanwhile the general good was conveniently forgotten by one and all. And when they couldn't agree, fists or inkpots started flying in the chamber. We certainly put an end to that, and no one can blame us for it."

"Have you ever thought that in 1919 the German people were forced to learn how to govern themselves for the first time, and that it was not all that easy for them to respect the rights of others, once those in power stopped using their authority to ensure justice and fair treatment?"

"That may well be so, but the parties have had fourteen years to learn just that, and things have been getting worse, not better. If we Germans keep fighting and lying to each other at home, we cannot really be surprised to find that the world outside should be fast losing what little respect it still has for us, and tries constantly to put us down. The League of Nations keeps talking about self-determination, but no one bothers to ask the people of South Tyrol whether they wish to remain part of Italy. And when they prate about security and disarmament, what they really mean is the disarmament of Germany and the security of other nations. You can't really blame our young men for refusing to swallow all these lies; indeed, you should be happy that they don't."

"And do you really believe that Hitler is any more honest?"

"I can see that Hitler is much too coarse for your liking. But he is speaking to ordinary people, and has to use their language. I can't prove that he is more honest than the rest, but you will soon realize that he is much more successful. You will see that Germany's old adversaries will make many more concessions to Hitler than they did to his predecessors, simply because, from now on, they, too, will have to make sacrifices if they intend to continue on their unjust course. In the past, things were very much easier for them, seeing that the German Government yielded to the least pressure from the outside."

"Even if you were right about that, I would not like to call a forced retreat by others a genuine achievement of your movement or of Hitler. Every concession she forces upon others will earn Germany fresh enemies, and the war ought to have taught all of us the folly of the slogan 'More foes, more honor.'"

"And so you think Germany should remain a nation despised and derided by all, a nation that takes every insult lying down, and accepts sole responsibility for the last war, simply because the others say so, or rather because, when all is said and done, we lost?"

"We don't seem to be speaking the same language," I said, trying to calm him. "Perhaps you will let me explain what I really think. To begin with, I have found that countries like Denmark, Sweden or Switzerland do quite well even though they have won no wars in the past hundred years and though they lack powerful armies. They are able to preserve their own character despite their semidependence on the great powers. Why shouldn't we be striving for the same thing? You may object that we are a much bigger and economically much stronger nation than the Swedes or the Swiss, and that our influence abroad ought to be correspondingly greater. But I am trying to look further ahead. The changes we are witnessing throughout the world are reminiscent of the changes that occurred in Europe at the end of the Middle Ages. At that time, technical advances, particularly in the manufacture of weapons, caused castle and town to lose their political independence, to make way for larger units, for territorial states, great or small. Once this change had taken place, the building of costly walls and defensive moats became a positive disadvantage—small towns which had dispensed with them could spread more easily and more quickly than those which continued to huddle behind great boulders. In our day, too, technology is making enormous strides, and defense techniques have been radically changed by the invention of the airplane. Consequently, there is once again the tendency to form larger political units, to go beyond national boundaries. Hence we would probably do more for national security if we disarmed and tried, instead, to establish friendly relations with our neighbors through economic contracts. If we rearmed, we should probably encourage others to do likewise, and the result would be a decrease in general security. Joining a wider political community might well afford us much better protection. I am only mentioning these facts to stress how difficult it is to assess the value of distant political objectives. I am firmly convinced that we must never judge political movements by their aims, no matter how loudly proclaimed or how sincerely upheld, but only by the means they use to realize these aims. Now when it comes to means, you National Socialists are no different from the Communists; the leaders of both movements have clearly lost faith in the persuasive force of their own ideas. Hence both leave me quite unmoved except for the fact that I am sadly convinced that both will bring down misfortune on Germany."

"But you have to admit that nothing at all has been achieved by what you call 'good means.' The Youth Movement organized no demonstrations, smashed no windows and did not beat up its opponents. It only tried, by personal example, to set new and better standards. But what did it achieve?"

"Perhaps nothing in political terms. But the Youth Movement does have some cultural achievements to its credit. You have only to think of public education, of crafts, of the Dessau Bauhaus, of the revival of old music, of song circles and amateur plays. Don't you think that is worth something, too?"

"Yes, perhaps. I don't wish to belittle any of it, and I am truly thankful for all that was done. But Germany also needs political liberation from her state of inner corruption, and no amount of goodwill has been able to achieve that. Surely, this cannot mean we must now sit by and do nothing. You criticize us because we follow a man whom you consider too coarse and of whose methods you disapprove. I, too, think that anti-Semitism is one of the most unpleasant aspects of our movement, and I hope that it will soon stop. But have any of the old school, any of the old professors who now complain about our revolution, ever tried to show us young people a better way, one that would have led us to our goal along a better road? Not one of them told us how to shake off our misery. You, too, kept silent. What else could we have done in the circumstances?"

"And so you resorted to force and made a revolution—in the mistaken belief that good can come out of destruction. Do you know what Jakob Burckhardt said about the consequences of revolution? 'How fortunate the revolution that does not end with the enthronement of the archenemy.' Why should we Germans have this extraordinary fortune? The reason why we older ones—I have to count myself among them now—gave you no advice was simply that we had no advice to offer—beyond the very humdrum counsel that everyone ought to do his work as conscientiously and decently as he can, hoping that his example may produce some good in the end."

"In other words, all you want is to preserve the old, to cling to the past. You disapprove of all changes, whereas the young are desperately anxious to forge ahead. If you had your way, nothing new could ever happen. And yet you see fit to propound revolutionary ideas in your own science. After all, relativity and quantum theory represent radical breaks with everything that has gone before."

"If we are discussing revolutions in science, we ought to look more closely at what has been happening. Take Planck's quantum theory. No doubt, you know that when Planck first tackled the subject he had no desire to change classical physics in any serious way. He simply wanted to solve a particular problem, namely, the distribution of energy in the spectrum of a black body. He tried to do so in conformity with all the established physical laws, and it took him many years to realize that this was impossible. Only at that stage did he put forward a hypothesis that did not fit into the framework of classical physics, and even then he tried to fill the breach he had made in the old physics with additional assumptions. That proved impossible, and the consequences of Planck's hypothesis finally led to a radical reconstruction of all physics. But even after that those realms of physics that can be described with the concepts of classical physics remained quite unchanged. In other words, only those revolutions in science will prove fruitful and beneficial whose instigators try to change as little as possible and limit themselves to the solution of a particular and clearly defined problem. Any attempt to make a clean sweep of everything or to change things quite arbitrarily leads to utter confusion. In science only a crazed fanatic—for instance, the kind of man who maintains that he can invent a perpetual-motion machine—would try to overthrow everything, and, needless to say, all such attempts are completely abortive. True, I don't know whether scientific revolutions can be compared with social revolutions, but I suspect that even historically the most durable and beneficial revolutions have been the ones designed to serve clearly defined problems and which left the rest strictly alone. Think of that great revolution two thousand years ago, whose maker said: 'Think not that I am come to destroy the law . . . but to fulfill it.' So let me repeat: what matters is to confine oneself to a single, important objective and to change as little of the rest as possible. The small part we have to change may well have so great a transforming force that it may affect all forms of life without any further effort on our part."

"But why do you cling so resolutely to the old forms? Surely, it happens only too often that they are out of step with the times and merely survive through a kind of inertia. Why not shake them off? For instance, I find it absurd that our professors should still appear at official functions dressed in their medieval gowns. Surely, that is one old vestige that ought to be discarded."

"Needless to say, I am not so much concerned with the old forms themselves as with what they stand for. Let me explain this, too, by an example taken from physics. The formulae of classical physics represent an old store of empirical knowledge that was not only correct in the past but will be correct at any time in the future. Quantum theory has merely given this store of knowledge a different form. But as far as the content is concerned there is nothing in the motion of pendulums, the lever laws, the motions of the planets, that needs to be changed, because the world itself does not change with respect to these processes. And this brings me to the problem of the gown: this old form of dress probably goes back to the time when people were divided into guilds, and it probably reflects the very much older recognition that men of learning are especially important to mankind, simply because their advice is the soundest. The gown is meant to symbolize this special position, and its wearer, even when he falls short of the standards of his guild, is protected against rude attacks by the common people. This experience is certainly as true today as it was a hundred years ago; but I agree that it matters very little whether or not we express it outwardly through the wearing of gowns. In any case, I suspect that many of those who object to the wearing of academic dress are also anxious to get rid of the meaning these gowns were designed to represent. That would be unspeakably silly of them; after all, the facts cannot be changed."

"You are again extolling experience as opposed to the rashness of youth, as old people are so accustomed to do. And since we can't argue back, we simply withdraw deeper into our shells."

My visitor now made as if to go, but I asked him whether I might not play him the last movement of the Schumann concerto—as far as this could be done at all without the help of an orchestra. He seemed happy to stay on, and when he finally took his leave, I felt that we were parting on good terms.

During the weeks following this conversation, political interference in university life became more and more intolerable. One of my faculty colleagues, the mathematician Levy, who, by law, should have enjoyed immunity because of his distinguished war record, was suddenly relieved of his post. The indignation of some of the younger members of the staff—I am thinking particularly of Friedrich Hund, Karl Friedrich Bonhoeffer and the mathematician B. L. van der Waerden—was so great that we thought of tendering our resignations and of persuading other colleagues to follow suit. Before taking this grave step, I decided to discuss the whole question with an older man, one who enjoyed our full confidence. I accordingly asked Max Planck for an interview and then paid a visit to his home in the Grunewald section of Berlin.

Planck received me in a somewhat somber but otherwise friendly and old-fashioned living room; all that was missing to complete the picture was an old oil lamp over its central table. Planck seemed to have grown a good many years older since our last meeting. His finely chiseled face had developed deep creases, his smile seemed tortured, and he was looking terribly tired.

"You have come to get my advice on political questions," he said right off, "but I am afraid I can no longer advise you. I see no hope of stopping the catastrophe that is about to engulf all our universities, indeed our whole country. Before you tell me about Leipzig—and, believe me, things couldn't be worse than they are here in Berlin—I would like to apprise you of my conversation with Hitler a few days ago. I had hoped to convince him that he was doing enormous damage to the German universities, and particularly to physical research, by expelling our Jewish colleagues; to show him how senseless and utterly immoral it was to victimize men who have always thought of themselves as Germans, and who had offered up their lives for Germany like everyone else. But I failed to make myself understood—or, worse, there is simply no language in which one can talk to such men. He has lost all contact with reality. What others say to him is at best an annoying interruption, which he immediately drowns by incessant repetitions of the same old phrases about the decay of healthy intellectual life during the past fourteen years, about the need to stop the rot even at this late hour, and so on. All the time, one has the fatal impression that he believes all the nonsense he pours forth, and that he indulges his own delusions by ignoring all outside influences. He is so possessed by his so-called ideas that he is no longer open to argument. A man like that can only lead Germany into disaster."

I now told him about the latest developments in Leipzig and about the plan of some of the younger staff members to resign. But Planck was convinced that all such protests had become utterly futile.

"I am glad to see that you are still optimistic enough to believe you can stop the rot by such actions. Unfortunately, you greatly overestimate the influence of the university or of academicians. The public would hear next to nothing about your resignation. The papers would either fail to report it or else treat your protests as the actions of misguided and unpatriotic cranks. You simply cannot stop a landslide once it has started. How many people it will destroy, how many human lives it will swallow up, is a matter of natural law, even if we ourselves cannot predict its precise course. Hitler, too, can no longer determine the subsequent course of events; he is a man driven by his obsessions and not someone in the driver's seat. He cannot tell whether the forces he has unleashed will raise him up or smash him to pieces.

"In these circumstances, your resignation would have no effect at the present time other than to ruin your own career—I know you are prepared to pay that price. But as far as Germany is concerned, your actions will only begin to matter again after the end of the present catastrophic phase. It is to the future that all of us must now look. If you resign, then, at best, you may be able to get a job abroad. What might happen at worst, I would rather not say. But abroad you will be one of countless emigrants in need of a job, and who knows but that you would deprive another, in much greater need than yourself? No doubt, you would be able to work in peace, you would be out of danger, and after the catastrophe you could always return to Germany—with a clear conscience and the happy knowledge that you never compromised with Germany's gravedigger. But before that happens many years will have passed; you will have changed and so will the people of Germany, and I don't know whether you will be able to adapt yourself to the new circumstances, or how much you will achieve in this changed world.

"If you do not resign and stay on, you will have a task of quite a different kind. You cannot stop the catastrophe, and in order to survive you will be forced to make compromise after compromise. But you can try to band together with others and form islands of constancy. You can gather young people around you, teach them to become good scientists and thus help them to preserve the old values. Of course, no one can tell how many such islands will survive the catastrophe; but I am certain that if we can guide even small groups of talented and right-minded young people through these horrible times, we shall have done a great deal to ensure Germany's resuscitation after the end. For such groups can constitute so many seed crystals from which new forms of life can arise. I am thinking first and foremost of the revival of scientific research in Germany. But since no one knows what precise roles science and technology will play in the future, these remarks may apply to much wider fields of endeavor as well. I think that all of us who have a job to do and who are not absolutely forced to emigrate for racial or other reasons must try to stay on and lay the foundations for a better life once the present nightmare is over. To do so will certainly be extremely difficult and dangerous, and the compromises you will have to make will later be held against you, and quite rightly so. Naturally, I cannot blame anyone who decides differently, who finds life in Germany intolerable, who cannot remain while injustices are committed that he can do nothing to prevent. But in the ghastly situation in which Germany now finds herself, no one can act decently. Every decision we make involves us in injustices of one kind òr another. In the final analysis, all of us are left to our own devices. There is no sense in giving advice or in accepting it. Hence I can only say this to you: No matter what you do, there is little hope that you can prevent minor disasters until this major disaster is over. But please think of the time that will follow the end."

And that is how we left it. On the train journey back to Leipzig, the conversation kept going round and round in my head. I almost envied those of my friends whose life in Germany had been made so impossible that they simply had to leave. They had been the victims of injustice and would have to suffer great material hardships, but at least they had been spared the agonizing choice of whether or not they ought to stay on. I tried again and again to pose the problem in different ways, to look at it from all angles. If a member of one's family catches a fatal infection, is it better to leave the house before one catches the infection and perhaps spreads it, or is it better to look after the patient even if he is bound to die? But then could one really compare a revolution with a disease? Might that not be a cheap way of suspending all moral judgments? And what precisely were the compromises Planck had hinted at? At the beginning of each lecture you had to raise your hand and give the Nazi salute. But hadn't I raised my hand to wave at acquaintances even before the advent of Hitler? Was that really a dishonorable compromise? And then you had to sign all official letters with "Heil Hitler." That was much less pleasant, but luckily I, for one, didn't have to write all that many official letters, and when I did, the new salutation invariably meant: "I don't want to have closer contact with you." We were expected to attend celebrations and marches, but I felt it ought to be possible to get out of quite a few. A compromise here, a compromise there, and where did you draw the line? Had William Tell been right to refuse homage to Gessler's hat, thus endangering the life of his own child? Ought he to have compromised? And if the answer was no, ought we to compromise with our own Gesslers?

Conversely, if one decided to emigrate, how could one reconcile that decision with Kant's dictum: "Act only on that maxim whereby thou canst at the same time will that it should become a universal law"? After all, not everyone could emigrate. Ought one to roam restlessly from one country to the next, in an attempt to avoid all possible social catastrophes? And when all was said and done, you belonged to a particular country by birth, language and education. And if you cut off your roots and moved, might you not be simply leaving the field to those madmen, those spiritually unhinged creatures whose demented plans were driving Germany headlong into disaster?

Planck had said that we might be faced with alternatives that would be equally unjust. Were such situations possible? I tried to think up an extreme situation which, though it had not occurred in reality, was not too farfetched nor quite obviously beyond a humane solution. This was the example I finally hit upon: A dictatorial government has jailed ten of its opponents and has decided to kill at least the most important of the prisoners. At the same time, the government is terribly anxious to justify this murder before the rest of the world. Accordingly, it makes an offer to another of its opponents, say, a jurist who has been left at liberty because of his high international renown: if he can produce and sign a legal justification for the murder of the most important prisoner, then the other nine will be released and allowed to emigrate. If he refuses, all ten prisoners will be killed. The jurist is left in no doubt that the dictator is in earnest. What is he to do? Is a clear conscience, a "white waistcoat," as we used to call it cynically, worth more than the lives of nine friends?

Even his suicide would be no solution; it would merely lead to the immediate slaying of the innocent ten.

Thinking along these lines, I remembered a conversation with Niels Bohr, during which he referred to the fact that justice and love were complementary concepts. Although both are essential components of our behavior toward others, they are, in fact, mutually exclusive. Justice would force the jurist to withhold his signature, the more so as the political consequences of his signing might be such as to destroy many more innocent people than the nine friends. But would love refuse the cry for help sent up by the desperate families of the nine friends?

After a while, I realized how extremely childish it was to go on playing such absurd mental games. What mattered was to decide here and now whether I ought to emigrate or to stay in Germany. "Think of the time after the catastrophe," Planck had said, and I felt he was right. We would have to form islands, gather young people round us and help them to live through it all, to build a new and better world after the holocaust. And this was bound to involve compromises, for which we would rightly be held to account—and perhaps even worse. But at least it was a worthwhile job. The world outside did not need us; there were others who could fulfill the tasks set there much better than we could. By the time the train pulled into Leipzig, I had made up my mind: I would stay on in Germany, at least for a time, continue working at the university, and, for the rest, do my bit as best as I possibly could.

13

Atomic Power and
Elementary Particles (1935–1937)

Despite the convulsion of scientific life at home and abroad
caused by Hitler's rise to power, atomic physics developed with
astonishing rapidity. In Lord Rutherford's Cambridge laboratory, Cockcroft and Walton succeeded in building a linear
accelerator capable of producing a beam of high-energy protons
(ionized hydrogen). With it they bombarded atoms of boron
and lithium, and so great was the energy of the accelerated
protons that they overcame the barrier of electrical repulsion, hit
the atomic nucleus and transformed it. By means of this and
similar accelerators, notably the cyclotron developed in America,
it was possible to make countless new experiments in nuclear
physics, so that a fairly clear picture of the properties of atomic
nuclei and the forces working within them could be built up
soon afterward. It appeared that atomic nuclei, unlike atoms,
could not be likened to small-scale planetary systems in which
lighter bodies revolve about a heavy central body; rather must
the atomic nuclei be considered as various-sized drops of the same
nuclear material, itself made up of protons and neutrons in
about the same proportions. The density of this material was
roughly the same for all atomic nuclei; however, strong electrostatic repulsion between protons ensured that in heavy nuclei
there were slightly more neutrons than protons. The assumption
that the powerful forces binding the nuclear material together
are invariant under the exchange of protons and neutrons was
proved correct, and the symmetry between protons and neutrons,

which I had envisaged so long ago in our skiing hut, was further borne out by the fact that while some atomic nuclei emit electrons during beta disintegration others emit positrons. So as to get a more detailed idea of the structure of the nucleus, we in Leipzig proceeded on the assumption that the nucleus was a nearly spherical drop of nuclear material, in which neutrons and protons moved about freely without appreciably interacting with, or disturbing, one another. Niels Bohr, in Copenhagen, on the other hand, thought that neutron-proton interactions were of great importance, and accordingly likened the nucleus to a kind of sandbag.

It was in order to resolve these differences that I went to Copenhagen for a few weeks some time between the autumn of 1935 and the autumn of 1936. As Niels' guest, I was assigned a room in the official residence which the Danish Government and the Carlsberg Foundation had put at the disposal of the Bohr family. For many years this house was a most important meeting center for atomic physicists. It was built in Pompeian style, and reflected the strong influence of the famous sculptor, Thorvaldsen. From the living room a broad flight of steps, flanked by statues, led into the park, with a central fountain among flower beds and massive old trees offering shelter from sun and rain. The vestibule of the house gave onto a conservatory, where the splashing of another small fountain was the only sound to break the otherwise perfect silence. Here we would watch Ping-Pong balls riding on the jet of the fountain and discuss the physical causes of this phenomenon. Behind the conservatory was a large hall with Doric columns which often served for receptions during scientific congresses. It was in this magnificent house that I now joined the Bohr family, and it so happened that Lord Rutherford, the father of modern atomic physics, as he was later called, was also spending part of his holidays there. Hence it was only natural that all three of us should from time to time walk through the park, discussing the latest experiments or the structure of the atomic nucleus. I shall now try to reconstruct one of these conversations.

Lord Rutherford: "What do you think would happen if we built even bigger accelerators or high-tension generators and fired protons of still greater energy and velocity at even heavier atomic nuclei? Will the projectile simply pass through the atomic nuclei, possibly without causing much damage, or will it remain stuck in them and surrender all its kinetic energy? If the interactions between individual nuclear particles are as important as Niels believes them to be, then the projectiles will get caught inside, but if protons and neutrons move about more or less freely in the atomic nucleus, i.e., without strong interactions, then the projectiles might easily pass through a nucleus without causing any great disturbance inside."

Niels: "I am certain that the projectiles will, as a rule, remain stuck in the atomic nucleus and that their kinetic energy will be more or less evenly distributed among all the nuclear particles, for the interactions are very great indeed. As a result of its collision with a projectile the atomic nucleus will simply grow hotter, and the rise in temperature can be calculated from the specific heat of the nuclear material and the energy of the projectile. What happens afterward might perhaps best be called a partial evaporation of the atomic nucleus. In other words, a few particles on the surface will occasionally receive so much energy that they will leave the atomic nucleus. But what do you think?"

The question was put to me.

"I am inclined to agree with you on the whole," I replied, "although your scheme differs considerably from our Leipzig model, according to which the particles move about with almost complete freedom inside the nucleus. A very quick-moving particle penetrating the nucleus would certainly be involved in several collisions because of the intensity of the interactions and thus lose its energy. Things may be quite different with a slow-moving particle, for its wave nature must come into play and the number of possible energy transfers become smaller. In that case the interaction may become unimportant. But it ought to be possible to work it all out by simple calculations—we know enough about the atomic nucleus for that. I shall make this my first task when I get back to Leipzig.

"May I now put a question to you? Do you think it likely that with more powerful accelerators we may one day be able to use nuclear energy for technical purposes—for instance, for the artificial creation of new chemical elements in appreciable quantities—or to utilize the energy of nuclear bonds much as we exploit the energy of chemical bonds during combustion? I believe there is an English science fiction story in which a physicist solves all his country's political difficulties, at home and abroad, by producing an atom bomb as a kind of *deus ex machina*. All that is, of course, nothing but wishful thinking, but in a somewhat more serious vein the physical chemist Walther Nernst once said in Berlin that the earth was a kind of powder keg, and that it needed only a match to blow it sky-high. And he was right: if only we could keep combining four hydrogen nuclei from the sea into one helium nucleus, we would free such enormous quantities of energy that Nernst's powder-keg comparison would become a ridiculous understatement."

Niels: "No one has really thought the whole matter through to its conclusion. The decisive difference between chemistry and nuclear physics is this: whereas most chemical experiments involve the majority of the molecules of the substances concerned—for instance, in the powder of your keg—in nuclear physics we can never experiment with more than a very small number of atomic nuclei. On principle, nothing in this can change with the invention of ever-greater accelerators. After all, the number of processes involved in a chemical experiment is to the number of nuclear processes so far produced by experimental techniques as, say, the diameter of our planetary system is to the diameter of a pebble, and this relationship is not materially altered if instead of the pebble we take a piece of rock. Of course, things would be quite different if we could raise a piece of matter to so high a temperature that the energy of the individual particles became great enough to overcome the repulsive forces between the atomic nuclei, and if, at the same time, we could keep the density high enough to ensure that collisions did not become too rare. But this calls for temperatures of something like a thousand million degrees, and long before we reached such temperatures the vessels in which we enclosed our experimental substances would have evaporated."

Lord Rutherford: "In any case, no one has seriously suggested that energy can be derived from nuclear processes. For though the fusion of a proton or neutron with an atomic nucleus does release energy, a much greater amount is needed to produce the fusion in the first place, for instance, by the acceleration of a very large number of protons, most of which will miss their target. The largest proportion of this energy is in any case dissipated in the form of Brownian movement. As far as the liberation of energy is concerned, experiments with atomic nuclei may therefore be called a sheer waste. All those who speak of the technical exploitation of nuclear energy are talking moonshine."

On this point we quickly reached agreement, and none of us suspected that, only a few years later, Otto Hahn's discovery of uranium fission would dramatically change the picture.

Of the prevailing political turmoil very little filtered through into the stillness of Bohr's park. We sat down on a bench in the shadow of the large trees and watched an occasional gust of wind deflect the descending spray of the fountain, causing individual droplets to settle on the rose leaves, where they glistened brightly in the sun.

After my return to Leipzig, I produced the promised calculations. They confirmed Niels' suspicion that quick-moving protons would generally remain stuck in the atomic nucleus, simply heating the latter by the collision. (It was at about the same time that processes of this type were actually observed with fast-moving protons from cosmic rays.) However, my calculations also seemed to suggest that the strong interactions of individual particles could, in a first approximation, be neglected during the study of the inner structure of atomic nuclei. We accordingly continued to work along these lines. Carl Friedrich, who was then assistant to Lise Meitner in Otto Hahn's Institute in the Dahlem section of Berlin, would often attend our siminars in Leipzig and, on one occasion, gave us a report of his own investigations of nuclear processes in the sun and the stars. He was able to adduce a convincing theoretical proof that certain well-defined reactions between light atomic nuclei take place in the innermost part of the stars, and that the enormous energy that is constantly emitted by the stars is obviously due to these nuclear processes. Hans Bethe published similar findings in America, and we became used to the idea of treating stars as gigantic atomic furnaces, in which atomic energy is released before our very eyes, admittedly not as a controllable process but as a natural phenomenon. But even then none of us seriously thought of the possibility of using atomic power for technological purposes.

In our Leipzig seminar we did not work only on the theory of atomic nuclei. Ever since my stay in the skiing hut, I had been devoting a great deal of fresh thought to the theoretical problems connected with the behavior of elementary particles. Paul Dirac's ideas on antimatter had since become an experimentally proven constituent of atomic physics. We knew that there was at least one process in nature during which energy is transformed into matter: the energy of radiation can give rise to electron-positron pairs. It seemed reasonable to assume that there would be other processes of the same kind, and we tried to imagine the nature of such processes during collisions of elementary particles traveling at very high velocities.

The next person with whom I discussed these problems was Hans Euler, who had joined us a few years earlier as a young student. I had noticed him almost immediately, not because he was of above-average intelligence but because of his striking appearance. He looked more delicate and far more sensitive than most of my other students, and his eyes bore the marks of suffering, particularly when he smiled. He had an elongated, almost sunken face, framed in curly fair hair, and when he spoke, one could feel a degree of concentration that was quite unusual in so young a man. It was easy to see that he was living in great poverty, and I was most happy to get him the modest job of lab assistant. It was not until very much later, however, when I had earned his trust, that he confided all his troubles to me. His parents could barely pay his fees. He himself was a convinced Communist; his father, too, was probably in trouble for political reasons. He was engaged to a young girl who had had to flee Germany because she was of Jewish descent, and now lived in Switzerland. Of those who had seized power in Germany since 1933 he could only think with disgust, and he would rarely speak of them. To relieve his financial burden, I would invite young Euler for lunch whenever I could, and on one such occasion I suggested that he might perhaps like to emigrate. But Euler would not hear of it; his ties with Germany were too close for that. But that was another subject on which he preferred to keep his own counsel.

And so we would talk about atomic physics instead and, in particular, about the possible consequences of Dirac's discovery, and the transformation of energy into matter.

"Dirac has shown," Euler said, "that when a light quantum flies past an atomic nucleus, it may change into a pair of particles —an electron and a positron. Does this mean the light quantum itself consists of an electron and a positron? In that case, it would be a kind of double star, one in which the electron and positron revolve about each other. Or is that a false picture?"

"I don't think it's very convincing. You see, the mass of a double star cannot be much smaller than the sum of the masses of its constituent parts. Nor would it necessarily have to move through space with the velocity of light. There is no reason why it should never come to rest."

"But what *can* we say about the light quantum in this context?"

"Perhaps that it is virtually made up of an electron and a positron. The word 'virtually' means that we are dealing with a possibility. In that case, my assertion means no more than that the light quantum may, in certain experiments, split up into an electron and a positron—nothing more."

"Well, in a very high-energy impact, a light quantum might easily be transformed into two electrons and two positrons. Does that mean that it is virtually made up of these four particles as well?"

"Yes, I believe that would be the consistent view. Since the term 'virtually' denotes possibilities, we are entitled to say that the light quantum is virtually made up of two or four particles. Two different possibilities do not necessarily exclude each other."

"But what is the advantage of this sort of assertion?" Euler asked. "We might equally well say that every elementary particle is virtually made up of any number of other particles. After all, any number of particles might be created during high-energy collisions. In that case our statement says very little indeed."

"I should not put it like that, for, you see, the number and type of particles are not as arbitrary as all that. Only such configurations may be considered possible descriptions of a particular particle as have the same symmetry as the original particle. Instead of 'symmetry,' we might say more precisely: transformation characteristics under operations that leave the physical laws unchanged. After all, quantum mechanics has taught us that the stationary states of an atom are characterized by their symmetries. Things are probably much the same with elementary particles, which, when all is said and done, are simply stationary states of matter."

Euler was still not fully satisfied. "The whole argument is a bit too abstract for my liking. What we probably ought to be doing is to think up experiments that would lead to unexpected results, and this precisely because light quanta are virtually made up of pairs of particles. It seems reasonable to assume that we should obtain at least qualitatively satisfactory results if we stuck to the model of the double star, and asked what conclusions orthodox physics would draw. For instance, we could investigate whether or not two light rays crossing in empty space really pass through each other with no interaction, as we have assumed until now, and as the old Maxwellian equations demand. If pairs of electrons and positrons are virtually present, i.e., contained as a possibility, in a light ray, then another light ray ought to be scattered by these particles; hence there would be deflection of light by light, that is, an interaction of the two light rays. We ought to be able to demonstrate its existence and to calculate its extent from Dirac's theory."

"Whether or not we would be able to observe it would, of course, depend on the intensity of the mutual perturbation. But by all means calculate the effect. Perhaps experimental physicists will then discover ways and means of corroborating your results."

"I really think this whole 'as if' philosophy is terribly odd. The light quantum is said to behave in some experiments as if it consisted of an electron and a positron. But at other times it apparently behaves as if it consisted of two or more such pairs. The result is a wishy-washy kind of physics. And yet we can use Dirac's theory to calculate the probability of a certain event with great precision, and find that experiments will confirm the results."

I tried to develop the "as if" approach a little further. "You know that experimental physicists have recently discovered yet another type of elementary particle, namely, the meson. Over and above that, there are the powerful forces which keep the atomic nucleus together and to which some elementary particles must correspond, reflecting the wave-particle dualism. Perhaps there are still a great many other elementary particles which we have missed, simply because they are too short-lived. We could then compare an elementary particle with an atomic nucleus or a molecule. That is, we could treat it 'as if' it were a collection of many, possibly different, elementary particles. In this connection, we might ask a question which Lord Rutherford recently put to me in Copenhagen in connection with atomic nuclei: 'What do you think would happen if we fired a high-energy elementary particle at another?' Will it remain stuck in the other elementary particle, considered as a cluster, heat it up and eventually cause its evaporation, or will it pass smoothly through the cluster without too much of a disturbance? Needless to say, the answer depends once again on the intensity of the interactions, and on this subject we know next to nothing. But perhaps it is worth our while to concentrate on the known interactions, and to see what happens with them."

At the time this conversation took place, the physics of elementary particles was still in its infancy. True, cosmic rays had provided physicists with certain experimental starting points, but systematic experimentation had not even been begun. Euler wanted to know what I thought of the future of this branch of atomic physics.

"Thanks to Dirac's discovery, i.e., the existence of antimatter," he said, "the whole picture has become much more complicated than ever it was. For a time it looked as if the whole universe was built up out of only three basic units: the proton, the electron and the light quantum. This was a simple enough picture, and there was good reason to hope that its essential features would soon be completed. Now the picture is getting increasingly confused. The elementary particle has ceased to be elementary; 'virtually' speaking at least, it is a very complicated structure. Doesn't that mean that we are that much further from true understanding?"

"No, I don't really think so. After all, the earlier picture with its three elementary units was not particularly convincing. Why ever should there have been just these three arbitrary units, and why should one of them, the proton, be precisely 1,836 times heavier than the other, the electron? What is so distinctive about the number 1,836? And why should these units have been indestructible? After all, we can shoot them at one another with tremendous force, so why should they hold together? Now, thanks to Dirac's discovery, things look much more reasonable than they did. The elementary particle, like the stationary state of an atom, is determined by its symmetry. The stability of forms, which Bohr made the starting point of his theory and which can be interpreted, at least in principle, by quantum mechanics, is also responsible for the existence and stability of elementary particles. These forms are always recreated if they are destroyed, just like the atoms of the chemists; and this is the natural consequence of the fact that symmetry is rooted in nature herself. Admittedly, we are still a long way from being able to formulate the physical laws responsible for the structure of elementary particles, but I could very well imagine that sooner or later they will lead us to even that strange number, 1,836. I am quite fascinated by the idea that symmetry should be something much more fundamental than the particle itself. This fits in with the spirit of quantum theory as Bohr has always conceived it. It also fits into Plato's philosophy, but this need not detain us now. Let us stick to what we can investigate. I think you ought to determine the scattering of light by light, while I look at the general question of what happens during the collision of high-energy particles."

This was the program to which both of us kept during the next few months. My own calculations showed that at high energies even the interaction involved in radioactive beta decompositions can become very marked, so that it seemed possible that a collision between two high-energy elementary particles could lead to the creation of many new particles. Now, at the time, we already had hints of this multiple-creation process in cosmic radiation, but still lacked reliable experimental corroboration—this was to come twenty years later, with the construction of the great accelerators. Meanwhile Euler, together with another of my pupils, B. Kockel, determined the scattering of light by light, and although no direct experimental verification was possible here, there is little doubt today that the scattering effect they deduced is a fact.

14

Individual Behavior in the Face of
Political Disaster (1937–1941)

The immediate prewar years, or rather what part of them I spent in Germany, struck me as a period of unspeakable loneliness. The Nazi regime had become so firmly entrenched that there was no longer the slightest hope of a change from within. At the same time, Germany became increasingly isolated, and it was obvious that resistance abroad was gathering momentum. A gigantic arms race had started, and it seemed only a question of time before the two camps clashed in open battle, a battle in which international law, Geneva conventions and moral inhibitions would all go completely by the board. In Germany itself this situation was aggravated further by the isolation of the individual. Communication became increasingly difficult—only the most intimate friends dared to speak their minds to one another; otherwise you resorted to the kind of language that hid far more than it revealed. I found life in this stifling atmosphere of distrust quite unbearable, and the certainty that it was all bound to lead to the total destruction of Germany only drove home to me the severity of the task I had set myself on returning from Max Planck.

I remember a gray, cold morning in January 1937, when I had to sell "Winter Aid" flags in the center of Leipzig. This activity, too, was part of the many humiliations and compromises we had to put up with at the time—although we could, of course, tell ourselves that collecting money for the poor was nothing to be ashamed of. Just the same, I was in a state of complete despair as I rattled my box, not because the show of subordination I had been forced to make bothered me in itself, but simply because of the utter senselessness and hopelessness of what I was doing and of what was happening all around me. Suddenly, I was in the throes of a strange and disturbing mental state. The houses in these narrow streets seemed very far away and almost unreal, as if they had already been destroyed and only their pictures remained behind; people seemed transparent, their bodies having, so to speak, abandoned the material world so that only their spirits remained. Behind these ghostly figures and the gray sky, I sensed a strong brightness. I noticed that several people stepped up to me with unusual cordiality, and gave me their "Winter Aid" contributions with looks that brought me out of my reveries and, for a moment, bound me closely to them. But then I was far away again, and began to fear that so much loneliness might well prove more than I could bear.

That evening I was asked to play chamber music at the Bückings. My host, a publisher, was a cellist; Jacobi, a jurist from the University of Leipzig and a dear friend, was an excellent violinist, and together we intended to play the Beethoven G Major Trio, which I knew so well from my youth—in 1920 I had played in the slow movement during the matriculation celebrations in Munich. But this time, in my delicate state of mind, I was afraid of the music and of meeting new people, and so I was delighted to see that our audience was a small one. But one of the young guests, on her first visit to the Bückings, managed to reach across to me even during our first conversation, and drew me back from the far reaches to which I had withdrawn. I felt I was on solid ground once again, and this sensation grew steadily stronger as I continued our conversation while playing the trio. We were married a few months later, and in the coming years Elisabeth Schumacher was to share all my difficulties and dangers with great fortitude and courage.

In the summer of 1937, I ran briefly into political trouble. It was my first trial, but I shall pass over it, because many of my friends had to suffer so much worse.

Hans Euler had become a regular visitor in our house, and we often discussed the political problems we were facing. On one occasion, Euler told me that he had been asked to attend a Nazi camp for lecturers and teachers in a nearby village. I advised him to go lest he endanger his university position, and told him about the Hitler Youth leader who had once poured out his heart to me and whom he would probably see there. Perhaps they would both benefit from the meeting.

When Euler came back from the camp, he was disturbed and agitated. He told us at length about his experiences.

"The people attending the camp are a very strange mixture. Many simply go because they are expected to, and because, like me, they are afraid to lose their jobs. I had very little contact with these. But then there is a smaller group, including your Hitler Youth leader, who really believe in National Socialism and are certain that good will come of it. Now, I know how much harm the Nazis have already done and what disasters Germany can expect from them in the future. But at the same time I feel that many of these young men want much the same things as I do myself. They, too, find this rigid, bourgeois life quite intolerable, detest a society in which material wealth and status matter above everything else. They want to replace this hollow sham with something richer, more vital. They want to render human relationships more human, and so, in fact, do I. I still fail to understand why these efforts must lead to such inhuman results. I just see that they do. And that makes me doubt my own beliefs as well. I always hoped for a Communist victory. If it had come, some who are up today would have been down and vice versa, and we would undoubtedly have done many things much better, but whether the sum total of man's inhumanity would have grown less I can no longer tell. Good intentions are obviously not enough. Stronger forces are brought into play, and these quickly get out of control. But surely that can't mean we must put up with the old at all costs, even though it is nothing better than a hollow sham. I honestly don't know what to think or what to do any more."

"We shall simply have to wait," I said, "until such time as we can do anything at all. Meanwhile we must try to keep order in those small corners to which our own lives are confined."

In the summer of 1938, the dark clouds on the international horizon had become so threatening that they began to cast a shadow even over my new home life. I had to do two months' service with the Mountain Rifle Brigade in Sonthofen, and on

many occasions we were ordered to stand by for immediate transfer to the Czech border. Then the clouds moved away once more,
but I was convinced that it could not be for long.

Toward the end of the year, something quite unexpected
happened in atomic physics. One Tuesday Carl Friedrich came
over from Berlin to inform the members of our Leipzig seminar
of Otto Hahn's discovery that barium was one of the products of
the bombardment of uranium atoms with neutrons. This meant
that the nucleus of the uranium atom had been split into
two comparable parts, and we at once asked ourselves whether
this process could be explained in terms of what we then knew
about the atomic nucleus. For a long time we had likened the
atomic nucleus to a droplet of protons and neutrons, and Carl
Friedrich had, years earlier, produced estimates of the energy,
surface tension and electrostatic repulsion inside the droplet,
based on empirical data. To our pleasant surprise, it now turned
out that the quite unexpected process of nuclear fission was
altogether in keeping with our general ideas. In very heavy
atomic nuclei this process could occur spontaneously and hence it
could be triggered off by a slight external action, for instance, by
bombardment with a neutron. It seemed almost incredible that
we should not have thought of this possibility earlier. A further
exciting consequence was that, immediately after the division,
the two parts of the divided nucleus would probably not be
perfect spheres. This meant that they would contain extra energy
which might later lead to some evaporation, that is, to the
emission of a few neutrons from the surface. Now these neutrons
might well hit other uranium nuclei, causing them to divide in
turn and so set off a chain reaction. Needless to say, a great deal
of experimental work would still have to be done before such
speculations became part and parcel of modern physics, but the
many new possibilities they opened up were enough to excite us
tremendously. Only a year later we came face to face with the
problem of the technical exploitation of atomic energy in peace
and war.

If we have to sail a ship into the storm, we close all the hatches,
take in sail, fasten anything that moves and look to our safety.
That was why, in the spring of 1939, I went in search of a house
in the mountains in which my wife and children could take

refuge from the coming disaster. I eventually found just the right place in Urfeld, above Lake Walchen, a few hundred feet up from the road on which Wolfgang Pauli, Otto Laporte and I had gone cycling so many years ago, discussing quantum theory while looking across the Karwendel Mountains. The house had belonged to the painter Lovis Corinth, and I had admired the view from the terrace at exhibitions.

I took yet another important step that year. I had many friends in America, and felt the need to see them before the war started—who knew if we would ever meet again? And I also realized that, if I was to help in Germany's reconstruction after the collapse, I would badly need their help.

In the summer of 1939 I lectured at the universities of Ann Arbor and Chicago. I used the opportunity to call on Enrico Fermi, with whom I had attended Born's seminars in Göttingen. For many years, Fermi had been Italy's leading physicist, but he had subsequently decided to ride out the coming storm in America. When I visited him in his home, he asked me whether it would not be better if I, too, made my home in the States.

"Whatever makes you stay on in Germany?" he asked. "You can't possibly prevent the war, and you will have to do, and take the responsibility for, things which you will hate to do or to be responsible for. If so much anguish might produce the least bit of good, then your remaining there might be understandable. But the chances of that happening are extremely remote. Here you could make a completely fresh start. You see, this whole country has been built up by Europeans, by people who fled their homes because they could not stand the petty restrictions, continuous quarrels and recriminations among small nations, the repression, liberation and revolution and all the misery that goes with it. Here, in a larger and freer country, they could live without being weighed down by the heavy ballast of their historical past. In Italy I was a great man; here I am once again a young physicist, and that is incomparably more exciting. Why don't you cast off all that ballast, too, and start anew? In America you can play your part in the great advance of science. Why renounce so much happiness?"

"I know just how you feel, and I have told myself the same thing thousands of times. Indeed, the idea of leaving the confines of Europe for the expanses of the New World has been a constant temptation ever since my first visit ten years ago. Perhaps I ought to have emigrated then. But instead I decided to collect a small circle of young people around me, people who wish to participate in the advances of modern science and who are anxious to make certain that uncontaminated science can make a comeback in Germany after the war. If I abandoned them now, I would feel like a traitor. The young, after all, cannot emigrate as easily as we can—they would have a hard time finding jobs abroad, and I would feel it quite improper to take advantage of my greater experience. Let's just hope that the war will be a very brief one. During the autumn crisis when I was a conscript, I noticed that only a handful of our people are really in favor of war. It is quite possible that, once the complete hypocrisy of Hitler's so-called peace policy becomes plain, the German people will make short shrift of Hitler and his followers. But I admit that it doesn't look like that at the moment."

"There is another problem," Fermi said, "that you cannot ignore. You know that Otto Hahn's discovery of atomic fission may be used to produce a chain reaction. In other words, there is now a real chance that atom bombs may be built. Once war is declared, both sides will perhaps do their utmost to hasten this development, and atomic physicists will be expected by their respective governments to devote all their energies to building the new weapons."

"That danger is, of course, real enough," I replied, "and you are only too right in what you say about our participation and responsibility. But is emigration really the answer? In any case, I have the certain feeling that atomic developments will be rather slow however hard governments clamor for them; I believe that the war will be over long before the first atom bomb is built. Of course, no one can look into the future, but major technical developments usually take quite a few years, and the war will certainly be finished before then."

"Don't you think it possible that Hitler may win the war?" asked Fermi.

"No. Modern wars call for vast technological resources, and because Hitler has chosen to cut off Germany from the rest of the world, our technical potential has grown incomparably smaller than that of our probable opponents. This situation is so obvious that I sometimes have the vague hope it may even filter through to Hitler himself, and that he may have second thoughts about starting a war. But this is probably pure wishful thinking on my part. For Hitler is irrational and simply shuts his eyes to anything he does not want to see."

"And you still want to return to Germany?"

"I don't think I have much choice in the matter. I firmly believe that one must be consistent. Every one of us is born into a certain environment, has a native language and specific thought patterns, and if he has not cut himself off from this environment very early in life, he will feel most at home and do his best work in that environment. Now history teaches us that, sooner or later, every country is shaken by revolutions and wars; and whole populations obviously cannot migrate every time there is a threat of such upheavals. People must learn to prevent catastrophes, not to run away from them. Perhaps we ought even to insist that everyone brave what storms there are in his own country, because in that way we might encourage people to stop the rot before it can spread. But then that might be going too far in the other direction. For, try as he may, the individual can often do nothing whatever to prevent the great mass of people from taking the wrong path. Under the circumstances, it would be wrong to expect him to sink or swim with those who have scorned his advice. In short, there are no general guidelines to which we can cling. We have to decide for ourselves and cannot tell in advance whether we are doing right or wrong. Probably a bit of both. Now I myself decided a few years ago to stay on in Germany— and even if my decision was wrong, I believe I ought to stick to it. For I knew even then that there would be a great deal of injustice and misfortune."

"That's a great pity," Fermi said. "Let's just hope we will meet again after the war."

Before leaving New York, I had a similar conversation with G. B. Pegram, an experimental physicist from Columbia University, who was older and more experienced than I, and whose advice meant a great deal to me. I was most grateful for the obvious concern with which he advised me to emigrate to America, but rather unhappy about my failure to get him to see my point of view. He found it quite incomprehensible that anyone in his right senses should wish to return to a country of whose imminent defeat he was firmly convinced.

The *Europa,* which took me back home early in August 1939, was almost empty, and its very emptiness served to underline all the arguments Fermi and Pegram had used.

We spent the second half of August in Urfeld, getting our new country house ready. Then, early on September 1, when I went down to the post office, the landlord of the hotel Zur Post ran up to me and said excitedly: "Do you know that war against Poland has started?" When he saw my horrified face, he added by way of consolation: "Don't worry, Herr Professor, it will all be over and done with in three weeks' time."

A few days later, I received my call-up papers. Quite unexpectedly I was ordered to report, not to the Mountain Rifles, with whom I had done my training, but to the Army Ordnance Department in Berlin. There I, like several of my colleagues, was told to work on the technical exploitation of atomic energy. Carl Friedrich had been given similar orders, which meant that we would have ample opportunity to meet in Berlin and to discuss our respective positions and attitudes. I shall try to sum up our conclusions in retrospect, as if they had been reached during a single conversation.

"So you, too, are a member of our uranium club," I may have begun, "and must have thought a great deal about the task we have been set. First of all, we are, of course, working in a very interesting branch of physics, and if we were at peace and nothing else were involved, we should probably be very happy to work at a project of such wide scope. But we happen to be at war, and everything we do may cause untold harm to others. So we had best watch our step very carefully."

"You are absolutely right, of course, and I have already thought of ways and means of getting out of this trap in one way or another. It ought not to prove too difficult to volunteer for front-line service or perhaps to work in a less alarming field. But in the end I decided to stick to our uranium problem, precisely because it has such vast possibilities. If the technical exploitation of atomic energy is still a very, very long way off, then it can do no harm to work in what you call the 'uranium club.' Indeed, doing so gives us a chance to protect all those talented young people whom we have been able to interest in atomic physics during the last decade. Again, if atomic technology is, so to speak, knocking at the gate, then it is far better to have some influence over developments than leave it all to others or to pure chance. Of course, we can't tell how long we as scientists can keep control, but there is likely to be quite a long intermediate stage in which physicists must have the last word."

"I feel this would be possible only if there were some kind of trust between the Ordnance Department and us. As it is, I was questioned by the Gestapo several times last year, and I hate to be reminded of the cellar in Prinz Albrecht Strasse, or of the ugly inscription painted on one of the walls: 'Breathe deeply and quietly.' "

"There can never be confidence or trust between officials, only between men. And why should there be no men of goodwill in the Ordnance Office, men who would meet us without prejudice and who would be ready and willing to talk things over with us? After all, it is in our common interest."

"Perhaps, but it's a very dangerous game just the same."

"There are many different degrees of trust. We may be able to get just enough to thwart the most irrational developments. But what do you think about the purely physical aspects of our problem?"

I tried to give Carl Friedrich a brief account of the very tentative theoretical studies I had begun during the first weeks of the war, and which amounted to little more than a physical tour of inspection.

"It looks very much as if no chain reaction can be triggered off by the bombardment of natural uranium with fast neutrons, in other words, that no atom bombs can be made with natural uranium," I began. "That's a bit of luck. A chain reaction can only be produced in pure or at least very strongly enriched uranium 235, and getting that—if it is possible at all calls for an altogether enormous technical effort. There may, of course, be other substances, but these are just as difficult to obtain. Atom bombs using these materials cannot possibly be built in the near future—neither by the English nor by the Americans nor by us. But if natural uranium is combined with a moderating substance which slows down all the neutrons liberated in the fission process, i.e., reduces them to the velocity of Brownian motion, then it may well be possible to start a chain reaction capable of yielding controllable amounts of energy. The moderator must not, of course, be allowed to capture too many neutrons. In other words, it must be a substance with a very small neutron absorption. Ordinary water will not do, but perhaps heavy water or pure carbon, possibly in the form of graphite, may be suitable. We shall have to test all this experimentally in the near future. I believe that we can work with a clear conscience—even in our relations with the officials—on chain reactions in this type of uranium pile and can leave the business of getting uranium 235 to others. The separation of isotopes, if it is possible at all, can only produce significant technical effects in the distant future."

"Do you really think that a uranium pile calls for much less technical effort than an atom bomb?"

"I am quite positive of that. The separation of two heavy isotopes, whose masses differ as little as those of uranium 235 and uranium 238, and their production in such quantities as will yield at least several kilograms of uranium 235, is a gigantic technical feat. In the uranium pile, on the other hand, all that is needed is a few tons of very pure natural uranium, together with graphite or heavy water. That calls for a very much smaller effort, differing by a factor of a hundred or perhaps as much as a thousand. I think your Kaiser Wilhelm Institute in Berlin and our group in Leipzig would do well to work on this problem."

"What you say seems reasonable enough," Carl Friedrich retorted, "and most reassuring, the more so as work on the uranium pile might prove very useful in the postwar period. If there is to be such a thing as a peaceful atomic technology, then it will have to be based on the uranium pile. Piles will provide energy for power stations, ships and the like, and we shall have done a good job by training a team capable of handling them.

"However, all of us must make a point in all our dealings with the Ordnance Department of saying little or nothing about the possibility of building atom bombs. Naturally, we shall have to keep this possibility constantly in mind, if only to be prepared for what the other side may have up their sleeves. However, I think it most unlikely, not least for historical reasons, that the outcome of the present war will be decided by atom bombs. So many irrational forces are at work, so many utopian hopes and so many bitter grudges, that if the issue were really settled with atom bombs, the outcome would be less satisfactory than one based on genuine understanding or even on sheer exhaustion. But the postwar world might well be constructed under the aegis of atomic technology and similar technical advances."

"So you, too, are discounting any possibility that Hitler might win the war?" I asked.

"To be quite honest, I am of two minds. People whose political judgment I respect, my own father chief and foremost among them, do not believe that Hitler has the least chance of winning the war. My father has always looked upon Hitler as a fool and a criminal who is bound to come to a bad end, and he has never wavered in this belief. But if that is the whole truth, how can we possibly explain Hitler's successes so far? Hitler's liberal and conservative critics have completely failed to grasp one decisive factor: his hold over the minds of the masses. I don't understand it myself, but I can certainly feel it. He has often enough confounded all his critics with his successes, and—who knows?— perhaps he will do it again."

"No," I replied, "not if the power game is played to the bitter end. For the technical and military potential of the British and the Americans is incomparably greater than ours. There is, of course, a vague chance that the other side may be afraid to go to the bitter end, lest they create a power vacuum in Central Europe, but horror at the crimes of the National Socialist system, particularly in racial matters, will probably outweigh all such scruples. Of course, no one can say just when the war will be over. Perhaps I am underestimating the powers of resistance of the political machine Hitler has built up. But, in any case, in doing our work, we must concentrate on the postwar period."

"You may be right," Carl Friedrich said in the end. "It is quite possible that I have unconsciously fallen victim to wishful thinking. For while no one in his senses can hope for Hitler's victory, no German can wish for the complete defeat of his country with all the terrible consequences that would entail. Still, with Hitler in the saddle there can't even be a compromise peace. God knows how it will all end, though I agree with you that we must prepare ourselves for the postwar period. That much at least is certain."

Experimental work on the new project was begun soon afterward, both in Leipzig and in Berlin. I was mainly involved in experiments to determine the properties of heavy water, which Robert Döpel had prepared most meticulously in Leipzig, but I often went to Berlin to follow the work done by various old colleagues and friends at the Kaiser Wilhelm Institute for Physics in Dahlem. Chief among them, apart from Carl Friedrich, was Karl Wirtz.

It was a great disappointment to me that I could not persuade Hans Euler to work on the uranium project. Just before the war, while I was away in America, Euler had become a close friend of Grönblom, one of my senior pupils. Grönblom was a Finn, unusually healthy and strong-looking, with a ruddy complexion, full of optimism, and convinced that, when all was said and done, the world was a good place to live in and full of challenges. As the son of a leading Finnish industrialist he was perhaps surprised to find himself so strongly drawn to a staunch Communist, but since human qualities counted for much more with him than opinions or creeds, he took Euler as he found him, with the openness and unstinting warmth so typical of youth. When war broke out, Euler was sadly shaken to find that Stalin had allied himself with Hitler in the division of Poland. And a few months later, when Russian troops invaded Finland, and Grönblom joined his regiment to defend his country's independence, Euler was a changed person. He spoke little, and I felt that he was deliberately cutting himself off, not only from me, but from every one of his friends, in fact from the whole world.

Euler himself had not been called up so far, no doubt because of his poor health. I was afraid, however, that they might send for him yet, and I accordingly asked him one day whether I should not try to get him an official transfer to our "uranium club." To my surprise, he told me that he had volunteered for the Luftwaffe. Since he noticed my agitation, he began to explain his reasons.

"You know that I have not volunteered because I want Germany to win. First of all, I don't believe in that possibility, and second, a victory by Nazi Germany would be as great a disaster as a Russian victory over the Finns. The unbridled cynicism with which our rulers cast all their principles to the winds if only they can score the least advantage fills me with utter despair. I have not, of course, volunteered for a unit in which I might be asked to kill people. I hope to join the reconnaissance branch of the Luftwaffe; there I may get shot down myself but will never have to fire a gun or throw a bomb, and that is all right. And in this ocean of senselessness, I can't really see what good I would do by working with you on the exploitation of atomic energy."

"As for the present catastrophe, we can do nothing about it," I objected, "neither you nor I. But after the war life will have to go on, here, in Russia, in America, everywhere. Before then very many people will have died, good people and bad, innocent men and guilty. And the survivors will have to try to build a better world. It won't be a particularly good one either, and people will quickly realize that the war has solved few problems. But they will nevertheless try to avoid some of the worst mistakes and do a couple of things better. Wouldn't you like to help?"

"I am not blaming anyone who sets himself that task. Those who have always been willing to take the world as they have found it, however inadequate, who have always preferred the painstaking path of gradual reforms to revolution, may find that they have been right all along, and that it is to them that the task of building a new world will fall. But as far as I am concerned, things are rather different. I had hoped that Communism would lead to the construction of a truly fraternal society. I have been proved wrong, and that is precisely why I don't want to take the easy way out now, why I don't wish to be treated better than any of the innocent people who are being slaughtered on all the fronts, be it in Poland, in Finland or elsewhere. Here in the Leipzig Institute there are many who wear the Nazi badge and so bear a somewhat greater responsibility for this war than the rest of us. And yet they have been exempted from military service. I find this quite intolerable, and I, for one, wish to remain true to myself. If you want to turn the world into a melting pot, then you must be ready to throw yourself into it as well. You must see my point."

"Indeed, I do only too well. But let us stick to your metaphoric melting pot. There is no reason to think that the molten mass, once it solidifies again, will assume the forms you might wish it would. For the forces presiding over the solidification result from the wishes of all the people concerned, not just from our own."

"If I shared your hopes, I should act differently, believe me. As it is, the present strikes me as so utterly futile as to rob me of what little courage I may have had in the past. But I do admire your optimism."

Soon afterward Euler went on a training course to Vienna, and the tone of his letters, which at first were as somber as our conversation, gradually became more relaxed. A few months later, when I delivered a lecture in Vienna, Euler invited me to take some new wine with him in a country inn, up beyond Grinzing. He refused to discuss the war. As we were looking down on the city, an airplane suddenly flew overhead, only a few yards above our heads. Euler smiled. It belonged to his own squadron, and had come to salute us.

At the end of May 1941, Euler wrote me a letter from the south. His squadron was flying reconnaissance missions from Greece over Crete and the Aegean. The letter seemed happy and almost abandoned; past and future seemed no longer to matter:

After fourteen days in Greece, we have forgotten everything that happens outside the glorious South. We can't even tell what day of the week it is. We are quartered in the Bay of Eleusis, and, whenever we are off duty, we live a glorious life, what with the blue waves and the wonderful sun. We have acquired a sailboat as well, and have lots of fun picking up meat and oranges. All of us wish we could stay here forever. There is little enough time left to dream between the old marble columns, but while we remain here, beneath the mountains and near the waves, past and present seem to have become as one.

As I mused about the change in Hans Euler's life, my thoughts were forced back to my conversation with Niels on the Öresund, and to Schiller's poem, which Niels had quoted on that occasion:

> Laughing at fears, he casts away
> All traces of terror and sorrow.
> He mocks at Fate's contrarious play.
> Let her strike today or tomorrow!

And if by chance she should delay,
Again he will toast the glorious day.

A few weeks later, German troops crossed the Russian border. Euler's plane never returned from its first reconnaissance flight over the Sea of Azov. His friend Grönblom was killed a few months later.

15

Toward a
New Beginning (1941–1945)

Toward the end of 1941 our "uranium club" had, by and large, grasped the physical problems involved in the technical exploitation of atomic energy. We knew that naturally occurring uranium, and heavy water, could be used to build a nuclear reactor which would supply energy and yield a disintegration product of uranium 239, which, like uranium 235, could serve as an explosive. Previously, i.e., toward the end of 1939, I had suspected, for theoretical reasons, that carbon could be used as the moderator in the place of heavy water. However, a measurement of the absorptive power of carbon had erroneously led to too high a value. Since this measurement had been made in another well-known institute, we had not bothered to repeat it and so had abandoned the whole idea prematurely. As for the production of uranium 235, we knew of no feasible methods that could have yielded significant quantities in Germany under war conditions. In short, though we knew that atom bombs could now be produced, in principle and by what precise methods, we overestimated the technical effort involved. Hence we were happily able to give the authorities an absolutely honest account of the latest development, and yet feel certain that no serious attempt to construct atom bombs would be made in Germany—the technical effort needed to achieve what seemed a very distant goal appeared so tremendous that Hitler could not possibly have decided on it in the tense situation our country now faced.

Nevertheless, we all sensed that we had ventured onto highly dangerous ground, and I would occasionally have long discussions particularly with Carl Friedrich von Weizsäcker, Karl Wirtz, Johannes Jensen and Friedrich Houtermans as to whether we were doing the right thing. I can clearly remember one conversation with Carl Friedrich in my room in the Kaiser Wilhelm Institute for Physics in Dahlem. Jensen had just left us, and Carl Friedrich said something like this:

"At present, we don't have to worry about atom bombs, simply because the technical effort seems quite beyond our resources. But this could easily change. That being so, are we right to continue working here? And what may our friends in America be doing? Can they be heading full steam toward the atom bomb?"

I tried to put myself into their position.

"The psychological situation of American physicists, and particularly of those who have emigrated from Germany and who have been received so hospitably, is completely different from ours. They must all be firmly convinced that they are fighting for a just cause. But is the use of an atom bomb, by which hundreds of thousands of civilians will be killed instantly, warrantable even in defense of a just cause? Can we really apply the old maxim that the ends sanctify the means? In other words, are we entitled to build atom bombs for a good cause but not for a bad one? And if we take that view—which has unfortunately prevailed throughout history—who decides which cause is good and which bad? It's easy enough to see that Hitler's cause is a very bad one, but is the Americans' good in every respect? Must we not judge it, too, according to the means by which it is pursued? Of course, even the good fight invariably involves some bad means, but is there not a point beyond which we cannot go under any circumstances? During the last century people tried to set a limit to the use of evil means through pacts and conventions. But in the present war these conventions are probably being ignored by Hitler no less than by his opponents. All in all, I think we may take it that even American physicists are not too keen on building atom bombs. But they could, of course, be spurred on by the fear that we may be doing so."

"It might be a good thing," Carl Friedrich told me, "if you could discuss the whole subject with Niels in Copenhagen. It would mean a great deal to me if Niels were, for instance, to express the view that we are wrong and that we ought to stop working with uranium."

In the autumn of 1941, when we thought we had a fairly clear picture of the technical possibilities, we asked the German Embassy in Copenhagen to arrange a public lecture for me there. I think I arrived in Denmark in October 1941, and when I visited Niels in his home in Carlsberg, I did not broach the dangerous subject until we took our evening walk. Since I had reason to think that Niels was being watched by German agents, I spoke with the utmost circumspection. I hinted that it was now possible in principle to build atom bombs, but that a tremendous technological effort was needed, and that physicists ought perhaps to ask themselves whether they should work in this field at all. Unfortunately, as soon as I mentioned the mere possibility of making atom bombs, Niels became so horrified that he failed to take in the most important part of my report, namely, that an enormous technical effort was needed. Now this, to me, was so important precisely because it gave physicists the possibility of deciding whether or not the construction of atom bombs should be attempted. They could either advise their governments that atom bombs would come too late for use in the present war, and that work on them therefore detracted from the war effort, or else contend that, with the utmost exertions, it might just be possible to bring them into the conflict. Both views could be put forward with equal conviction, and, in fact, during the war it turned out that even in America, where conditions were incomparably more favorable for the attempt than in Germany, the atom bomb was not made ready before V-E Day.

Niels, as I have said, was so horrified by the very possibility of producing atomic weapons that he did not follow the rest of my remarks. Perhaps he was also too filled with justifiable bitterness at the brutal occupation of his country by German troops to entertain any hopes of international understanding among physicists. I found it most painful to see how complete was the isolation to which our policy had brought us Germans, and to realize how war can cut into even the most long-standing friendships, at least for a time.

Despite this failure of my mission to Copenhagen, the German "uranium club" was in a relatively simple situation. The government decided (in June 1942) that work on the reactor project must be continued, but only on a modest scale. No orders were given to build atom bombs, and none of us had cause to call for a different decision. As a result, our work helped to pave the way for a peaceful atomic technology in the postwar period, and as such it was to bear useful fruits, despite and after all the destruction. It was perhaps no accident that the nucleus of the first atomic power station sent abroad (to Argentina) by a German firm was based on natural uranium and heavy water, just as we had planned it should be during the war.

In this connection, I clearly remember a conversation that brought me into closer contact with Adolf Butenandt, at the time a biochemist in one of the Kaiser Wilhelm Institutes in Dahlem. Although both of us had participated in a series of lectures on biology and atomic physics, we did not have a lengthy talk until the evening of March 1, 1943, when after an air attack we walked back from the center of Berlin to Dahlem.

We had just attended a meeting of the Academy for Aeronautics in the Air Ministry, off the Potsdamer Platz. Hubert Schardin had been lecturing on the physiological effects of modern bombs, mentioning, among other things, that the sudden build-up of air pressure due to an explosion in one's immediate vicinity might lead to a relatively painless death from an embolism. Toward the end of the meeting, the alert had been sounded and all of us had made for the Ministry shelter, fitted out with camp beds and paillasses. This was our first experience of very heavy bombing. Several bombs hit the building of the Ministry, we heard the collapse of walls and ceilings, and for a time we did not know whether the corridor between our shelter and the outside world was still open. The lights had gone out shortly after the start of the raid, and there were occasional gleams of a flashlight. A groaning woman was brought in and tended by two medical orderlies. At first we had all been talking and even laughing, but gradually we fell silent; the only sound then was the occasional thud as yet another bomb dropped nearby. After two particularly violent bursts, with pressure waves that shook the whole shelter, I heard Otto Hahn pipe up in a corner: "I bet Schardin doesn't believe in his own theories right now." With that, the atmosphere grew just a shade less somber.

When the raid was over, we managed to scramble out over a jumble of concrete block and twisted steel. Outside we were met by a fantastic view. The whole square in front of the Ministry was lit up by red flames from the upper floors of the surrounding buildings. Here and there, the fire had spread to the ground floor as well, and there were blazing pools of phosphorus in the middle of the street. The square was crowded with people anxious to get back to their homes, and vainly hoping for transport to the suburbs.

Butenandt and I had left the shelter together and decided to keep each other company while making for our homes in Fichteberg and Dahlem. At first we consoled ourselves with the thought that the raid might have been confined to the heart of the city and that the suburbs had been spared. Then we saw that the Potsdamer Strasse was flanked by garlands of flames for many miles ahead. In some places the fire brigade had set to work, but for all the good that did, they might as well have tried to empty a lake with a teaspoon.

We had to reckon on at least one and a half hours' brisk walking from the Potsdamer Platz to Dahlem, and so we had time for a fairly long conversation, not about Germany's present situation—that was only too obvious all around us—but about our hopes and plans for the postwar period.

Butenandt asked me: "What do you think are our chances of doing scientific research in Germany after the war? Many of our best institutes will have been destroyed, many young scientists will have been killed, and our people will be far too poor to put scientific development high on their list of priorities. Yet scientific research is probably a prerequisite of economic revival— without it Germany has little chance of taking her place in the European community."

"I think there is good reason to hope," I replied, "that many Germans will remember the work of reconstruction after the First World War, and that some of the most important contributions, for instance in the chemical or optical industries, resulted from the combined efforts of scientists and engineers. Our people will probably come to see quite quickly that modern life is impossible without fundamental research, and they will probably realize that Nazi neglect of science was one of the reasons for Germany's present collapse. Of course, that's by no means the whole story. The root of the evil lies considerably deeper. What we see before us is only the natural consequence of that myth of the twilight of the gods, of that 'all or nothing' philosophy to which the German people have time and again fallen prey. Their faith in a Führer, a hero destined to lead them out of danger and misery into a brighter and nobler future, free of all external constraint, or else, if fate should have decided against them, ready to march resolutely to their doom—this terrible creed is our greatest scourge. It replaces reality with a gigantic illusion, and prevents any real understanding between us and the nations with whom we have to live. And so I would prefer to put your question like this: Once our illusions have been completely and remorselessly shattered by reality, can scientific research help us Germans to arrive at a sober and critical view of the world and of our own position in it? In other words, I am thinking more of the educational than of the economic aspects of science, of its possible role in the development of critical thought. Of course, the number of people who can play an active part in science will never be very large, but scientists have always been highly respected in Germany; their counsels have generally been heeded, so their views may be expected to receive a fair hearing."

"Education for rational thought," Butenandt said, "is certainly a worthwhile task, and we must do our utmost to bring it about after the war. In fact, the way things have gone, people's eyes ought to have been opened to reality long since, for instance, to the fact that faith in the Führer is no substitute for raw materials, verbiage no viable alternative to scientific and technical achievements. A glance at the map, at the gigantic territories under the control of the United States, Great Britain and the Soviet Union, and at the tiny little area that is Germany, ought to have been enough to warn us against military adventurism. But we Germans find it extremely difficult to think logically and soberly. We are certainly not lacking in intelligent individuals, but as a nation we are inclined to be dreamers, to prize the imagination above the intellect, to exalt emotion above reason. Hence there is an urgent need to bring scientific thought back into honor, and that should not be too difficult during the unromantic times that are bound to follow this war."

We were still walking up the Potsdamer Strasse, and its continuation—Hauptstrasse, Rheinstrasse and Schlossstrasse—between rows of burning houses. Often we had to skirt piles of burning or red-hot timber, or improvised fences around unexploded bombs. At one point I walked into some liquid phosphorus and my right shoe began to burn. I quickly stepped into a puddle and so saved my precious footwear.

"We Germans," I tried to continue, "tend to look upon logic and the facts of nature—and even this debris all around us is nothing but natural fact—as a sort of straitjacket which we must wear, but only for lack of anything better. We think that freedom lies only where we can tear this jacket off—in fantasy and dreams, in the intoxication of surrender to some sort of utopia. There we hope, at long last, to realize the absolute whose existence we dimly suspect and which spurs us on to ever greater achievements, for instance in art. But we fail to appreciate what 'realization' means. Its very basis is reality; it can only be attained through the combination of facts or thoughts in accordance with the laws of nature. But even making due allowance for our strange propensity for indulging in fantasy and mysticism, I really cannot see why so many of our compatriots should find the scientific approach dull and disappointing. It is a common mistake to think that all that matters in science is logic and the understanding and application of fixed laws. In fact, imagination plays a decisive role in science, and especially in natural science. For even though we can hope to get at the facts only after many sober and careful experiments, we can fit the facts themselves together only if we can feel rather than think our way into the phenomena.

"Perhaps we Germans, of all people, have a special part to play in this area precisely because the absolute exerts so strange a fascination on us. Abroad, pragmatism is far more widespread than it is here, and we need only look at our neighbors or at history—that of Egypt, Rome or the Anglo-Saxon world—to appreciate how successful this approach can prove in technology, economics or politics. But in science and art those philosophical principles which the ancient Greeks developed to such magnificent effect have proved more successful still. If Germany has made scientific or artistic contributions that have changed the world—we have only to think of Hegel and Marx, of Planck and Einstein, of Beethoven and Schubert—then it was thanks to this love of the absolute, thanks to the pursuit of principles to their ultimate consequences. But only when the hankering after the absolute is subordinated to appropriate forms—in science to logical thought; in music to the rules of harmony and counter-point—only then, only under this extreme constraint, can it reveal its full power. The moment we try to explode these forms, we produce the kind of chaos we can see all around us. And I myself am not prepared to glorify this chaos with such concepts as the twilight of the gods or Armageddon."

As I was speaking, my right shoe had caught fire again, and it took me quite some time to scrape off all the phosphorus.

Watching me, Butenandt said: "It wouldn't be a bad thing if we simply bothered about the facts that stare us in the face. As for the future, we can only hope that Germany will develop the kind of politicians who can add imagination to a sense of reality, and so create halfway decent conditions of life for the German people. As far as science is concerned, the Kaiser Wilhelm Society will probably make a good basis for the revival of German re-search. The universities, after all, have been far less successful in escaping political interference than the Society, and will accord-ingly run into much greater trouble. For though our Society has also been forced to compromise, for instance by collaborating in military projects, many members have nevertheless kept up friendly relations with foreign scientists, with men who appreci-ate the importance of sober, reflective thought in Germany no less than in their own countries, and who might therefore be prepared to help us. Is that true in your branch of science as well? And what do you think of the chances of peaceful inter-national cooperation in it?"

"There will certainly be the peaceful exploitation of atomic energy, based on the method of uranium fission discovered by Otto Hahn. Since we have good reason to believe that no atom bombs will be built before this war is over—the technical effort involved is much too great—there is hope for fruitful interna-tional collaboration in the postwar period. After all, the decisive step was Hahn's discovery, and, when all is said and done, atomic physicists throughout the world have always worked peacefully together."

"Well, we shall just have to wait and see. In any case, we in the Kaiser Wilhelm Society will have to stick together."

We parted on this note, Butenandt making for Dahlem and I

for Fichteberg, where I had been staying with Elisabeth's parents for some time. I had recently brought my two oldest children to Berlin for their grandfather's birthday, and I was understandably anxious to find out how they and the old people had fared during the air raid. My hopes that Fichteberg might have been spared were quickly dispelled—from a distance I could see that the house flanking ours was ablaze, and that flames were leaping from our own roof as well. When I ran past the house next door, I heard cries for help, but I felt I had first to look after my own children and their grandparents. Our house had been badly hit; doors and shutters were blown in, and to my horror I discovered that there was no sign of life inside. It was only when I raced up into the attic that I spotted my wife's brave mother, wearing a steel helmet as protection against the falling rubble and fighting the flames with all her strength. She told me that the children had been taken in by the neighbors on the Botanical Garden side and that they were sleeping peacefully there in the care of their grandfather and Minister Schmidt-Ott and his wife. In our own house most of the flames had already been put out, so that all I had to do was to pull down a few smoldering rafters.

Only then did I go to the aid of the burning house next door. Most of the roof had collapsed, and the garden was strewn with burning beams. The whole upper floor was ablaze. On the ground floor, I discovered the young woman who had been calling for help. She told me that her old father was still up in the attic fighting a losing battle against the flames with buckets of water which he kept filling from one of the few taps that was not yet stopped up. The staircase had collapsed, and she did not know how he could possibly be brought down. Luckily I had put on an old, tight-fitting track suit that allowed me maximum freedom of movement. I scaled the walls to the roof, where, behind a wall of fire, I could see a white-haired old gentleman, scattering water almost mindlessly as he tried to stand his ground in an ever-diminishing circle of flame. I leaped across to him, and I could see how completely taken aback he was by the sudden appearance of a complete stranger, blackened with soot from head to toe. He immediately put the bucket down, straightened his back, bowed politely and said: "My name is von Enslin; most kind of you to come to my aid." He had the characteristic Prussian attitude I had always admired: simplicity, discipline and few words. I suddenly remembered my walk with Niels on the shores of the Öresund, but I had no time to reflect on the power of ancient examples now—action was urgently needed. And, in fact, I did manage to get the old man down along the same route I had clambered up.

A few weeks later, I moved my family from exposed Leipzig to Urfeld, as I had planned to do just before the war. Our Kaiser Wilhelm Institute for Physics in Dahlem had also received orders to move from its vulnerable quarters to a textile factory in the small town of Hechingen in southern Württemberg.

Of the chaotic last years of the war I retain only too few clear memories. But since these became part of the background against which I later based my opinions on general political questions, I feel that I must mention them in brief.

Among the most enjoyable aspects of my life in Berlin were the meetings of the so-called Wednesday Society, whose members included General Ludwig Beck, Minister Johannes Popitz, the famous surgeon Ferdinand Sauerbruch, Ambassador Ulrich von Hassel, Eduard Spranger, Jens Jessen, Count von der Schulenburg and others. I remember one evening at Sauerbruch's, who, after delivering a lecture on pulmonary surgery, treated us to what, at the time, was a princely dinner with glorious wine, so that after the dessert von Hassel jumped onto the table and sang student songs. I also recall our last get-together in July 1944, at which I acted as host. In the afternoon I had been picking raspberries in the Institute gardens, and the management of Harnack House had contributed milk and a little wine to my frugal entertainment. I spoke to my guests about atomic energy in the stars and of its technical exploitation on earth—or rather about those aspects that were not on the top-secret list. Beck grasped immediately that all the old military ideas would have to be changed, and Spranger put into words what we physicists had been thinking for a long time, namely, that the development of atomic physics might cause far-reaching changes in man's social and philosophical attitudes.

On July 19, I took the minutes of that meeting to Popitz, and then boarded the night train to Munich and Kochel. From there, it was a two-hour walk to Urfeld. On the way, I met a soldier dragging his luggage up the Kesselberg in a handcart. I added my heavy suitcase and helped him pull. The soldier told me he had just heard on the radio that an attempt had been made on Hitler's life. Hitler himself had received slight injuries, but the army command in Berlin was now in open revolt. When I asked him cautiously how he felt about it all, he only said: "It's time something was done." A few hours later I was sitting in front of my radio, and heard that General Beck had been killed in Army HQ in Bendler Strasse. Popitz, Hassel, Schulenburg and Jessen were mentioned as his accomplices, and I knew what that meant. Adolf Reichwein, who had visited me in Harnack House early in July, had also been arrested.

A few days later I went to Hechingen, where I found most members of my Berlin Institute assembled. We prepared to set up our atomic reactor in a cave in picturesque little Haigerloch, underneath the rock on which the church was built. My regular bicycle trips between Hechingen and Haigerloch, the orchards and the woods, in which we went searching for mushrooms during holidays—all this brought the present as glowingly to life for us as the waves in the Bay of Eleusis must have done for Hans Euler. For whole days, we could forget past and future. In April 1945, when the fruit trees began to blossom, the war was nearing its end. I arranged with my colleagues that, as soon as the Institute and its members were out of immediate danger, I would leave Hechingen on my bicycle and join my family in Urfeld.

In the middle of April, the last German stragglers passed through Hechingen heading east. One afternoon we could hear the first enemy tanks. In the south, the French had probably advanced well past Hechingen, as far as the ridge of the Rauhe Alb. It was high time I was gone. Toward midnight, Carl Friedrich returned from a bicycle reconnaissance tour of Reutlingen. We held a brief farewell celebration in the air-raid shelter of the Institute, and at about 3 A.M. I set off in the direction of Urfeld. By dawn I had reached Gammertingen, and it seemed that I was well past the front line. The only threat now came from low-flying aircraft, and to avoid them I traveled mostly at night, resting and foraging for food while the sun was up. I remember a hill near Krugzell, where I went to sleep in the shelter of a hedge, after a short repast in the warm sun. Beneath the cloudless sky, the whole Alpine chain lay stretched out before me—Hochvogel, Madelegabel and all the peaks I had climbed as a mountain rifleman seven years earlier. Below, the cherry trees were now in full blossom. It was real spring, and as I sank into a deep sleep, my confused thoughts seemed filled with light and a new hope.

A few hours later, I was awakened by something like a thunderclap, and when I opened my eyes I could see thick clouds of smoke rising from distant Memmingen. The barracks had just been bombed out of existence. The war was not yet over, and I had to keep going east. And so it was not until three days after I had set out that I reached Urfeld, and found my family well and unharmed. We spent the next week dragging sandbags in front of the cellar windows and laying in what stores of food we could get hold of. All our neighbors had fled to the opposite shore of the lake. The forest was full of scattered Wehrmacht and SS units, and its floor was littered with abandoned guns and ammunition, from which the children had to be kept away. In the daytime, we had to be on our guard against stray bullets that were still being fired all around us, and at night, too, we felt most uncomfortable in our no-man's land. On May 4, when Colonel Pash, leading a small U.S. detachment, came to take me prisoner, I felt like an utterly exhausted swimmer setting foot on firm land.

Snow had fallen during the night, and as I left, the spring sun shone down upon us out of a dark blue sky, spreading its brilliant glow over the snowy landscape. When I asked one of my American captors, who had fought in many parts of the world, how he liked our mountain lake, he told me it was the most beautiful spot he had ever seen.

16

The Responsibility of the Scientist (1945–1950)

After brief stops in Heidelberg, Paris and Belgium my captors finally took me to Farm Hall, where I was reunited with a few old friends and young collaborators of the uranium club. They included Otto Hahn, Max von Laue, Walther Gerlach, Carl Friedrich von Weizsäcker and Karl Wirtz. Farm Hall lies at the edge of Godmanchester, some twenty-five miles from the old English university city of Cambridge, and I was familiar with the landscape from earlier visits to the Cavendish Laboratory. This time there were ten of us, and we all came to look upon Otto Hahn, whose attractive personality and quiet, reflective attitude in a difficult position we greatly admired, as our obvious spokesman. He would negotiate with our captors whenever it was necessary, and this was not very often; the officers in charge of us did their job with extraordinary tact and humanity, so that after only a short while our relationship became one of complete mutual trust. We had been asked very little about our atomic researches, and we thought it rather odd that they should take so little interest in our work and yet guard us so carefully and prevent us from making even the slightest contact with the outside world. When I asked whether the American and the British had also been studying the uranium problem, I was told by the American physicists who had been sent to interrogate us that, unlike us, Allied scientists had devoted all their attention to tasks connected with the immediate war effort. This seemed quite plausible, in view of the fact that throughout the war there had been not the slightest hint of American work on nuclear fission.

On the afternoon of August 6, 1945, Karl Wirtz suddenly rushed in to tell me about a special news flash: an atom bomb had been dropped over Hiroshima. At first I refused to believe it, for I was convinced that the construction of atom bombs involved enormous technical efforts and probably the expenditure of many thousands of millions of dollars. I also found it psychologically implausible that scientists whom I knew so well should have thrown their full weight behind such a project. Under the circumstances, I was much more inclined to believe the American physicists who had interrogated us than some radio announcer who had perhaps been ordered to broadcast some sort of propaganda story. Moreover, Wirtz had told me that the word "uranium" had not been mentioned in the bulletin; this seemed to suggest that if any bombs had been dropped, they could not have been "atom bombs" in the sense that I used that term. But later in the evening, when the newscaster described the gigantic technical efforts that had been made, I had reluctantly to accept the fact that the progress of atomic physics in which I had participated for twenty-five long years had now led to the death of more than a hundred thousand people.

Worst hit of all was Otto Hahn. Uranium fission, his most important scientific discovery, had been the crucial step on the road toward atomic power. And this step had now led to the horrible destruction of a large city and its population, of a host of unarmed and mostly innocent people. Hahn withdrew to his room, visibly shaken and deeply disturbed, and all of us were afraid that he might do himself some injury. That night we said many ill-considered things, and it was not until next morning that we managed to put some order into our confused thoughts.

Behind Farm Hall, an old red-brick building, was a somewhat neglected lawn on which we used to play fist ball. Between the lawn and the ivy-covered wall that was our boundary lay an elongated rose garden, tended chiefly by Gerlach. It was surrounded by a path which we used much as medieval monks must have used the cloister. It was just the place for serious tête-à-têtes. On the morning after the terrifying news Carl Friedrich and I walked up and down in it for hours, thinking and talking. We began by voicing our anxiety about Otto Hahn, and Carl Friedrich then expressed the thought that was oppressing all of us:

"It is easy to see why Hahn should be dejected. His greatest scientific discovery now bears the taint of unimaginable horror. But should he really be feeling guilty? Any more guilty than any of us others who have worked in atomic physics? Don't all of us bear part of the responsibility, a share of his guilt?"

"I don't think so," I told him. "The word 'guilt' does not really apply, even though all of us were links in the causal chain that led to this great tragedy. Otto Hahn and all of us have merely played our part in the development of modern science. This development is a vital process, on which mankind, or at least European man, embarked centuries ago—or, if you prefer to put it less strongly, which he accepted. We know from experience that it can lead to good or to evil. But all of us were convinced—and especially our nineteenth-century rationalist predecessors with their faith in progress—that with growing knowledge good would prevail and evil could be kept under control. The possibility of constructing atom bombs never seriously occurred to anyone before Hahn's discovery; nothing in physics at the time pointed in that direction. To have played a part in so vital a scientific endeavor cannot possibly be considered a form of guilt."

"There will, of course, be quite a few," Carl Friedrich remarked, "who will contend that science has gone far enough. They will argue that there are far more important social, economic and political tasks to be done. They may, of course, be right, but all those who think like them fail to grasp that, in the modern world, man's life has come to depend on the development of science. If we were to turn our backs on the continuous extension of knowledge, the number of people inhabiting the earth in the fairly near future would have to be cut down radically. And that could only be done by means as horrible as the atom bomb or perhaps even worse.

"And then knowledge is power. As long as power struggles continue on earth—and at the moment their end is not even in sight—we must also fight for knowledge. Perhaps one day we may have a world government—and let us hope that it will be as free as possible—under which the search for further scientific knowledge does not have to be quite so frantic. But that is not our problem today. For the present, the development of science is a vital need of all mankind, so that any individual contributing toward it cannot be called guilty. Our task, now as in the past, is to guide this development toward the right ends, to extend the benefits of knowledge to all mankind, not to prevent the development itself. Hence the correct question is: What can the individual scientist do to help in this task; what are the precise obligations of the scientific research worker?"

"If we look upon the development of science as an historical process on a world scale," I replied, "your question reminds me of the old problem of the role of the individual in history. It seems certain that in either field the individual is replaceable. If Einstein had not discovered relativity theory, it would have been discovered sooner or later by someone else, perhaps by Poincaré or Lorentz. If Hahn had not discovered uranium fission, perhaps Fermi or Joliot would have hit upon it a few years later. I don't think we detract from the great achievement of the individual if we express these views. For that very reason, the individual who makes a crucial discovery cannot be said to bear greater responsibility for its consequences than all the other individuals who might have made it. The pioneer has simply been placed in the right spot by history, and has done no more than perform the task he has been set. As a result, he may possibly be able to exert just a little extra influence on the subsequent progress of his discovery, but that is all. In fact, Hahn invariably made a point of speaking out in favor of the exclusive application of uranium fission to peaceful purposes; in Germany he was loud in his warnings and counsels against any attempts to use atomic energy in war. Of course, he had no influence on developments in America."

"What is more," Carl Friedrich continued, "we must probably make a clear distinction between the discoverer and the inventor. As a rule, the former cannot predict the practical consequences of his contribution before he actually makes it, the less so as many years may go by before it can be exploited. Thus Galvani and Volta could have had no conception of the subsequent course of electrical engineering, nor can the slightest responsibility be attached to them for the uses and abuses of subsequent developments. Inventors seem to be in quite a different position. They have a définite, practical goal in view, and ought to be able to judge its merits. Hence we can apparently hold them answerable for their contributions. Yet it is precisely the inventor who can be seen to act not so much on his own behalf as for society at large. The inventor of the telephone, for instance, knew that society was anxious to speed up communication. In much the same way the inventor of firearms may be said to have acted on the orders of a society desirous of increasing its military strength. Hence no more than partial blame can be attached to him either, the less so as neither he nor society can foresee all the consequences of his invention. A chemist, for instance, who discovers an agricultural pesticide can tell you no more than the farmer what the ultimate consequences will be in regard to changes in the insect population due to his intervention. In short, we can ask no more of the individual than that he should try to set his own objectives in a wider context, that he should not thoughtlessly endanger the many for the sake of the few. All we can really ask of the individual is that he pay careful and scrupulous attention to the wider framework into which all scientific and technical progress must fit, even when this does not seem to further his immediate interests."

"If you draw a line between invention and discovery, where precisely do you put the atom bomb, the most recent and terrifying product of technical progress?"

"Hahn's fission experiments were a discovery, the manufacture of the atom bomb an invention. The physicists who built the bomb in America were inventors; they were not acting on their own behalves but on the overt or implicit orders of a warring group anxious to obtain the maximum striking power for its army. You once said that, for purely psychological reasons, you could not imagine that American physicists would put their whole hearts into the production of the atom bomb. Only yesterday you were still reluctant to accept the truth of the Hiroshima story. What do you think of our colleagues in America now?"

"Perhaps U.S. physicists were afraid that Germany might be the first to produce atom bombs. And understandably so, for, after all, uranium fission was discovered by Hahn, and atomic physics had reached a very high standard in Germany before

Hitler drove out so many of our most capable physicists. A Nazi victory with the atom bomb must have seemed so ghastly a threat that anything seemed justified to stop it, even an atom bomb of one's own. I don't think any of us could really object to that, particularly if we consider what happened in the concentration camps. After the end of the war in Europe, no doubt, many American physicists advised against the use of this terrible weapon, but by that time they no longer had a decisive say. In this respect, too, we cannot really criticize them, for which one of us was able to prevent any of the revolting crimes our own government has committed? The fact that we did not know the full extent of these crimes is no excuse, for we ought to have made greater efforts to find out.

"The worst thing about it all is precisely the realization that it was all so unavoidable. Throughout history, people have acted on the principle that right must be defended by might. Or in more evil and blatant form: that the end justifies the means. And what alternative could we put up against that attitude?"

"We have already said," Carl Friedrich replied, "that we might expect the inventor to fit his aim into the wider context of technical progress on earth. Let us look at the implications. Immediately after major catastrophes, people tend to draw up rather rash balance sheets. This time they may say that the atom bomb helped to end the war more quickly, that there might have been many more victims had the carnage been allowed to continue more slowly. I think you mentioned this argument yourself last night. But all such calculations are quite unsatisfactory, for none of us can predict the political repercussions of the bomb. May not the bitterness caused by it pave the way for later wars that will demand even more sacrifices? Will the new weapons produce a change in the balance of power, which, once all the great powers own atom bombs, may have to be rectified at the cost of untold lives? No one can predict these developments, so all such balance sheets are a waste of time. Why not start from quite a different principle, one we have often discussed, namely, that it is the choice of the means which determines whether a cause is good or bad?"

"Scientific and technical progress will undoubtedly lead to the constant expansion of an ever-diminishing number of superpowers," I replied. "The result will be centralization on an unprecedented scale, and we can only hope that it will leave the individual and the individual nation some freedom of action. This sort of development seems quite unavoidable to me; the only question is whether or not many fresh disasters will happen before the world finally settles down to a more stable order. In any case, we may take it that those few superpowers that will remain after this war will try to extend their spheres of influence as far as possible. This they can only do by way of alliances based on common interests, similarities in their social systems or in the values to which they subscribe, or else by exerting economic or political pressure. Whenever a weaker country outside the immediate sphere of influence of a great power is threatened or oppressed by a stronger country, the great power is likely to intervene in favor of the weaker and so increase its own influence. That is how we ought to look upon U.S. intervention in the two world wars, and there is no reason to think that this trend will stop now, nor can I see why we should object to it.

"Of course, some will brand all great powers engaging in this type of expansionist policy as imperialists, but here more than elsewhere the choice of means seems to me the decisive criterion. A great power that does not wield the big stick but prefers normal economic and cultural methods in its foreign dealings, and avoids the least suspicion of wishing to interfere with the life of its neighbors, will be much less open to reproach than one that can be seen to use force. And the political system of that great power which eschews all forms of undue pressure is likely to become the model for the world of the future. Now many people have come to look upon the United States as a bastion of liberty, as having the kind of social system in which the individual can develop his personality most fully. The fact that Americans enjoy complete freedom of expression, value personal initiative, respect individual views, treat prisoners of war better than most other countries—all this and many other facts had given rise to high hopes that the American political system provided the rest of the world with just the model it needed. The American Government ought to have remembered this hope when it decided whether or not to order the dropping of an atom bomb over Japan. I fear these hopes have been struck a bitter blow by the use of that bomb. Now all America's rivals will raise the cry of 'imperialism,' and their voices are bound to carry some weight. Precisely because the atom bomb was no longer needed for victory, its use will be interpreted as a show of naked power, and it is indeed difficult to see how we can proceed from here to a genuinely free world order."

"In other words," Carl Friedrich repeated, "you do think that the technical possibilities of the atom bomb ought to have been viewed against the wider context, that is, as part of a universal process of scientific and technical development leading inexorably to the establishment of a unified world order. In that case, it would have been obvious to one and all that the use of the bomb at a time when victory was already assured was a step in the wrong direction, weakening confidence in America's good faith and casting doubt on America's world mission. The existence of an atom bomb as such is no disaster, though it will help to restrict complete political independence to a few large powers with gigantic economic reserves. The smaller states will lose some of their independence, but that does not necessarily mean a restriction of individual freedom and may be considered the price we have to pay for the general improvement of living conditions.

"However, we are straying from the real problem. We were wondering about the behavior of the individual interested in technical progress while living in a world of conflicting ideas, passions and delusions. Our ideas on this subject seem to be rather hazy."

"We are nevertheless agreed," I countered, "that the individual tackling a scientific or technical task, however important, must nevertheless try to think of the broader issues. And, indeed, if he did not, why did he exert himself in the first place? Moreover, he will arrive at the correct answer more readily, the more he bears the wider connections in mind."

"In that case, if he wants to act for the best and not just leave it at noble thoughts, he will probably have to play a more deliberate part in public life, try to have a greater say in public affairs. Perhaps we should welcome this trend, for inasmuch as scientific and technical advances serve the good of society, those responsible for them will be given a greater say than they currently enjoy. Obviously, this does not mean that physicists or technicians could make better political decisions than the politicians themselves. But their scientific work has taught them to be objective and factual, and, what is more important, to keep the wider context in view. Hence they may introduce a measure of logical precision into politics, of greater objectivity and of respect for the facts. If we believe that, then we cannot but blame American physicists for not having tried hard enough to make their voices heard in public and for leaving to others the decision to use the atom bomb long before they had to do so. For I have no doubt at all that the evil consequences of dropping the bomb must have been quite obvious to them from the start."

"I don't know whether the word 'blame' is appropriate in this context. I simply feel that in this particular respect we happened to be luckier than our friends across the Atlantic."

We were released in January 1946, and returned to Germany. At last we could throw ourselves into the work of reconstruction of which we had thought and spoken so much ever since 1933, but which, as it turned out, proved more arduous than we had expected. The Kaiser Wilhelm Society could not be resuscitated either in its old form or in its old home, partly because the political future of Berlin was too uncertain, and partly because the Allies disapproved of the name, or any other reminder of the hated Kaiser. However, the British were kind enough to make the buildings of the former Aerodynamic Research Institute in Göttingen available for our purposes, and so we moved to the city in which I had met Niels Bohr twenty years earlier, and where I had later studied under Born and Courant. Max Planck, now almost ninety years old, had also been taken to Göttingen at the end of the war, and now helped us to set up a successor to the Kaiser Wilhelm Society, i.e., a scientific body that could coordinate the work of research institutes, old and new. I was fortunate enough to find a house for my family in the immediate vicinity of Planck's residence, and he would often chat with me across the garden fence and occasionally come over to listen to chamber music.

In those days, much effort and work had to be expended to satisfy the most primitive personal needs, and, in the Institute, on getting even the simplest equipment. But they were happy days for all that. No longer were we told, as we had been throughout the preceding twelve years, that this or that was impossible; everything seemed possible once again, and we could feel, in both scientific and private life, that things were getting better almost month by month, as people worked side by side with joyful enthusiasm. The great help we were given by various representatives of the occupying power proved of more than purely material benefit; it gave us the chance to feel part of a larger community once again, of a community desirous of building a better world, shaped by reasonable hopes for the future rather than by regrets about the past.

This change in emphasis was driven home to me by two conversations that have remained fresh in my memory. One was my first postwar meeting with Niels Bohr in Copenhagen. The reason for my visit was rather absurd, and I only mention it here so as to convey some idea of the atmosphere in Göttingen in those summer months of 1947. The British Secret Service had been tipped off by a source unknown to us that the Russians were planning to kidnap Otto Hahn and me into the Russian Zone, just a few miles away. When the British had good reason to believe that the Russian agents had already moved in, they at once transferred Hahn and me to Herford, which was near the administrative center of the British Zone. There I learned that, while awaiting further developments, the occupying authorities had made plans for me to visit Niels in Denmark. I was further informed that Ronald Fraser, the British officer who was our friendly custodian in Göttingen, wanted to talk to Bohr and myself about my visit to Niels in October 1941. A British military plane took us from Bückeburg to Copenhagen airport, whence we continued by car to Niels' country house in Tisvilde. And there we were, sitting down once again before the same fireplace that had witnessed so many discussions on quantum theory, walking along the same sandy forest paths on which, twenty years earlier, we had ambled down to the beach holding Niels' children by the hand. But when we tried to reconstruct what had been said during our conversation in the autumn of 1941, we noticed that both our memories had become blurred. I was convinced that I had broached the critical subject during a nocturnal walk on Pileallé, while Niels seemed certain that I had done so in his study in Carlsberg. All Niels could remember was the fright my carefully chosen phrases had given him, but he had completely forgotten my reference to the nearly insuperable technical problems and my question of what he thought physicists ought to do in this situation. After a while, we both came to feel that it would be better to stop disturbing the spirits of the past.

As once upon a time in the Bavarian mountains, it was the progress of physics that turned our thoughts from the past to the future. C. F. Powell in England had just sent Niels photographs of the tracks of what appeared to be previously unidentified elementary particles. Powell had, in fact, discovered the π-meson, which has played so important a role in the physics of elementary particles ever since. We at once discussed the possible connection between it and the forces in the atomic nucleus and, since the lifetime of the new particle seemed to be shorter than that of all known ones, we thought it possible that there might be many others as well, which had simply escaped observation because they were too short-lived. We thus saw ourselves facing a vast field of interesting research, to which we and the young people we had gathered around us would be able to devote our attention for many years to come. I, for one, was determined to do just that in the Institute that was being built in Göttingen.

When I returned home, Elisabeth informed me that there had been some substance in the kidnaping story after all. Two Hamburg dock workers had been arrested outside my house one night. Someone had apparently promised them large sums of money if they carried me off to a motorcar parked in the vicinity. I was struck by the fact that the whole plot had been so amateurish. The reason was discovered by British Intelligence some six months later. A disgruntled ex-Nazi, unable to find a job, had hit upon the whole scheme as a means of insinuating himself into the good books of the Allies. It was he who had recruited the two dock workers only to inform the British of the impending "abduction." He took everyone in for a while, though, like most schemers, not for long. We often had occasion to laugh about the whole story.

The second conversation I remember brought home to me how urgent it was to turn from the past toward the future. After Max Planck's death, Otto Hahn became the leading spirit in rebuilding our organization, now christened the Max Planck Society, and later became its first president. At the time, I was helping the Göttingen physiologist Hermann Rein to set up a Research Council that would act as a link between the Federal administration and scientific research. It was quite obvious that scientific progress and the resulting technological developments would have an extraordinarily important effect not only on the reconstruction of our own cities and industries, but also on the social structure of Europe as a whole. But as I had told Butenandt during our walk through burning Berlin, I was not exclusively concerned with getting the maximum government aid for scientific research; I was equally anxious to ensure for science a wider degree of influence over government decisions. I was firmly convinced that those responsible for running the new Germany must constantly be reminded that their job called for more than the balancing of conflicting interest; that there were unpleasant realities rooted in the very structure of the modern world, escape from which into emotional thinking could only lead to catastrophe. In other words, I wished to procure for science some right to take the initiative in public affairs. Adenauer, with whom I had many talks on the matter, promised me his full support. At the same time, many university and local government officials were trying to resurrect the Emergency Society for German Science which had been led by Schmidt-Ott in the twenties and which had rendered inestimable services to German science after the First World War. I felt unhappy about this attempt to recapture the past—the old idea that the government ought to support scientific research financially but that otherwise the two must go their separate ways struck me as completely out of step with the age. On this subject, which gave rise to many heated discussions, I remember particularly my lengthy conversation with the jurist, Ludwig Raiser, who later became president of the Scientific Council and held that office for many years. When I contended that the Temporary Association he supported might help to foster the deep-rooted German tendency to shut out unpleasant realities and to withdraw to the safety of our own ivory towers, Raiser said quite simply: "The two of us can't possibly hope to change the German character." I felt at once that he was right, that it was rarely the intention of individuals, but only hard pressure from the outside, that changed the out-look of large groups. And in fact, despite Adenauer's support, my own plans came to nothing: I failed to win the university and educational authorities over to the new ideas. It took ten years before external pressures led to the creation of a Research Ministry, which, by appointing consultative committees, helped to implement at least part of the suggestions I and so many like-minded colleagues had put forward. The Max Planck Society was much more easily brought into step with the needs of the modern world. Beyond that, our only consolation was that the much-needed educational reforms might yet be introduced, although only in the wake of protracted struggles and disputes.

17

Positivism, Metaphysics
and Religion (1952)

The resumption of international contacts once again brought together old friends. Thus, in the early summer of 1952, atomic physicists assembled in Copenhagen to discuss the construction of a European accelerator. I was most interested in this project because I was hoping that a large accelerator would help us to determine whether or not the high-energy collision of two elementary particles could lead to the production of a host of further particles, as I had assumed; whether, indeed, we were entitled to assume the existence of many new particles and, if so, whether, like the stationary states of atoms or molecules, they differed only in their symmetries, masses and lifetimes. The main topic of the meeting was thus a matter of great personal concern, and if I do not report it here, it is simply because I must relate a conversation with Niels and Wolfgang on that occasion. Wolfgang had come over from Zurich, and the three of us were sitting in the small conservatory that ran from Bohr's official residence down to the park. We were discussing the old theme, namely, whether our interpretation of quantum theory in this very spot, twenty-five years ago, had been correct, and whether or not our ideas had since become part of the intellectual stock-in-trade of all physicists. Niels had this to say:

"Some time ago there was a meeting of philosophers, most of them positivists, here in Copenhagen, during which members of the Vienna Circle played a prominent part. I was asked to address them on the interpretation of quantum theory. After my lecture, no one raised any objections or asked any embarrassing questions, but I must say this very fact proved a terrible disappointment to me. For those who are not shocked when they first come across quantum theory cannot possibly have understood it. Probably I spoke so badly that no one knew what I was talking about."

Wolfgang objected: "The fault need not necessarily have been yours. It is part and parcel of the positivist creed that facts must be taken for granted, sight unseen, so to speak. As far as I remember, Wittgenstein says: 'The world is everything that is the case.' 'The world is the totality of facts, not of things.' Now if you start from that premise, you are bound to welcome any theory representative of the 'case.' The positivists have gathered that quantum mechanics describes atomic phenomena correctly, and so they have no cause for complaint. What else we have had to add—complementarity, interference of probabilities, uncertainty relations, separation of subject and object, etc.—strikes them as just so many embellishments, mere relapses into prescientific thought, bits of idle chatter that do not have to be taken seriously. Perhaps this attitude is logically defensible, but, if it is, I for one can no longer tell what we mean when we say we have understood nature."

"The positivists would probably claim," I remarked, "that 'understanding' is tantamount to 'predictive ability.' If we can predict just a few special events, we have merely understood a small segment of nature, but if we can predict a large range of events, our understanding is correspondingly greater. There is a continuous scale from understanding very little to understanding almost everything, but there is no qualitative difference between predictive ability and understanding."

"Do you yourself believe there is such a difference?"

"Yes, I am convinced of it," I replied, "and I think we discussed it all some thirty years ago, on our bicycle tour round Lake Walchensee. Let me use an analogy. If we see an airplane in the sky, we can predict with a limited degree of certainty where it will be a second later. We shall assume either that it will continue in a straight line or, if it has begun to bank, that it will describe an arc. Though we shall be right most times, we still cannot claim that we have 'understood' the path. Only if we have spoken to the pilot beforehand and have learned from him how he intends to navigate can we claim to have understood his flight."

Niels was not entirely satisfied. "I think you may find it difficult to apply your analogy to physics. For my part, I can readily agree with the positivists about the things they want, but not about the things they reject. Let me explain. Their approach, as we know it particularly from England and America, and which is, in fact, little more than a systematization of earlier ideas, is rooted in the very ethos that presided over the dawn of modern science. Until then philosophers had concentrated on the great universal issues, analyzing them in the light of the teachings of the old authorities—chiefly Aristotle and Church doctrine. Experiential details were brushed aside as being unimportant; only the broader picture mattered. As a result, all sorts of superstitions crept in and philosophy made no progress: after all, the old authorities were dead. Early in the seventeenth century came emancipation from these masters, and a turning toward experience, that is, toward experimental studies of specific details.

"People tell us that, when such famous scientific institutes as the Royal Society in London were first founded, they tried to eradicate superstition by designing experiments to refute the claims of the most popular magic books. One such claim was that if a stag beetle was put in a chalk circle while certain magical formulae were recited at midnight, it would become spellbound. Scientists accordingly drew a circle on a table, recited the requisite formulae, placed the beetle into the center and let everyone see how quickly it escaped. In other famous academies, the members had to pledge that they would never discuss the great universals, but only specific facts. Theories about nature henceforth had to bear on individual groups of phenomena, not on their wider connections. A theoretical formula became a kind of guide to action—analogous to the notebook of the modern engineer, which contains a host of useful data on, say, the tensile strength of rods. Even Newton's well-known statement—'I do not know what I may appear to the world, but to myself I seem to have been only a boy playing on the seashore, and diverting myself in now and then finding a smoother pebble or a prettier shell than the ordinary, whilst the great ocean of truth lay all undiscovered before me'—even this statement expresses the ethos presiding over the dawn of modern science. Needless to say, Newton did very much more than this modest claim would suggest. He was able to express the fundamental laws governing a great many natural phenomena in mathematical terms. But this, he felt, was something 'whereof one must be silent.'

"It is quite understandable that, in their rebellion against authority and superstition, scientists should often have gone too far. There were, for instance, many old reports about stones falling out of the sky, and several monasteries and churches had even preserved such stones as relics. Now in the eighteenth century all these reports were dismissed as rank superstition, and the monasteries were asked to throw their worthless stones away. The French Academy even passed a special resolution to reject all further reports about stones dropping out of the sky, and it did not relent until a very large number of meteorites came down in the vicinity of Paris. I mention this fact merely to draw attention to the mental attitude typical of the dawn of modern science. There is no need to remind you how many new experiments and how much scientific progress have sprung from precisely this attitude.

"Now, all the positivists are trying to do is to provide the procedures of modern science with a philosophical basis, or, if you like, a justification. They point out that the notions of the earlier philosophies lack the precision of scientific concepts, and they think that many of the questions posed and discussed by conventional philosophers have no meaning at all, that they are pseudo problems and as such best ignored. Positivist insistence on conceptual clarity is, of course, something I fully endorse, but their prohibition of any discussion of the wider issues, simply because we lack clear-cut enough concepts in this realm, does not seem very useful to me—this same ban would prevent our understanding of quantum theory."

"When you say it would prevent our understanding of quantum theory," Wolfgang said, "do you mean that physics does not simply consist of experiment and mathematical formulae but that it must also philosophize where the two meet? In other words, that we must use everyday language to explain the precise interplay of experiment and mathematics? I myself have a strong

suspicion that all the difficulties of quantum theory will be found
to reside in this meeting, a fact most positivists choose to ignore,
precisely because their narrow concepts break down at this point.
The experimental physicist must be able to talk about his experiments and therefore he is forced to employ the concepts of
classical physics, although he realizes full well that they provide
an inadequate description of nature. This is his fundamental
dilemma, and one he cannot simply dismiss."

"Positivists," I tried to point out, "are extraordinarily prickly
about all problems having what they call a prescientific character. I remember a book by Philipp Frank on causality, in which
he dismisses a whole series of problems and formulations on the
ground that all of them are relics of the old metaphysics, vestiges
from the period of prescientific or animistic thought. For instance, he rejects the biological concepts of 'wholeness' and
'entelechy' as prescientific ideas and tries to prove that all
statements in which these concepts are commonly used have no
verifiable meaning. To him 'metaphysics' is a synonym for 'loose
thinking,' and hence a term of abuse."

"This sort of restriction of language doesn't seem very useful to
me either," Niels said. "You all know Schiller's poem, 'The
Sentences of Confucius,' which contains these memorable lines:
'The full mind is alone the clear, and truth dwells in the deeps.'
The full mind, in our case, is not only an abundance of experience but also an abundance of concepts by means of which we
can speak about our problems and about phenomena in general.
Only by using a whole variety of concepts when discussing the
strange relationship between the formal laws of quantum theory
and the observed phenomena, by lighting this relationship up
from all sides and bringing out its apparent contradictions, can
we hope to effect that change in our thought processes which is a
sine qua non of any true understanding of quantum theory.

"For instance, it is often said that quantum theory is unsatisfactory because, thanks to its complementary concepts of 'wave'
and 'particle,' it prohibits all but dualistic descriptions of nature.
Yet all those who have truly understood quantum theory would
never even dream of calling it dualistic. They look upon it as a
unified description of atomic phenomena, even though it has to
wear different faces when it is applied to experiment and so has

to be translated into everyday language. Quantum theory thus provides us with a striking illustration of the fact that we can fully understand a connection though we can only speak of it in images and parables. In this case, the images and parables are by and large the classical concepts, i.e., 'wave' and 'corpuscle.' They do not fully describe the real world and are, moreover, complementary in part, and hence contradictory. For all that, since we can only describe natural phenomena with our everyday language, we can only hope to grasp the real facts by means of these images.

"This is probably true of all general philosophical problems and particularly of metaphysics. We are forced to speak in images and parables which do not express precisely what we mean. Nor can we avoid occasional contradictions; nevertheless, the images help us to draw nearer to the real facts. Their existence no one should deny. 'Truth dwells in the deeps.' This claim is no less true than the first proposition of Schiller's poem.

"You mentioned Philipp Frank's book on causality. Philipp Frank was one of the philosophers to attend the congress in Copenhagen, and he gave a lecture in which he used the term 'metaphysics' simply as a swearword or at best as a euphemism for unscientific thought. After he had finished, I had to explain my own position, and this I did roughly as follows:

"I began by pointing out that I could see no reason why the prefix 'meta' should be reserved for logic and mathematics—Frank had spoken of metalogic and metamathematics—and why it was anathema in physics. The prefix, after all, merely suggests that we are asking further questions, i.e., questions bearing on the fundamental concepts of a particular discipline, and why ever should we not be able to ask such questions in physics? But I should start from the opposite end. Take the question 'What is an expert?' Many people will tell you that an expert is someone who knows a great deal about his subject. To this I would object that no one can ever know very much about any subject. I would much prefer the following definition: an expert is someone who knows some of the worst mistakes that can be made in his subject, and how to avoid them. Hence Philipp Frank ought to be called an expert on metaphysics, one who knows how to avoid some of its worst mistakes—I was not quite sure whether Frank was very happy about my praise, though I was certainly not offering it tongue in cheek. In all such discussions what matters most to me is that we do not simply talk the 'deeps in which the truth dwells' out of existence. That would mean taking a very superficial view."

That same evening Wolfgang and I continued the discussion alone. It was the season of the long nights. The air was balmy, twilight lasted until almost midnight, and as the sun traveled just beneath the horizon, it bathed the city in a subdued, bluish light. And so we decided to walk along the Langelinie, a beautiful harbor promenade, with freighters discharging their cargo on either side. In the south, the Langelinie begins roughly where Hans Christian Andersen's Little Mermaid rests on a rock beside the beach; in the north it is continued by a jetty that swings out into the basin and marks the entrance to Frihavn with a small beacon. After we had been looking at the toing and froing of the ships in the twilight for quite a while, Wolfgang asked me:

"Were you quite satisfied with Niels' remarks about the positivists? I gained the impression that you are even more critical of them than Niels himself, or rather that your criterion of truth differs radically from theirs. And I'm not quite sure to what extent Niels would be prepared to agree with you."

"I can't tell either, of course. After all, Niels grew up at a time in which it needed great strength of character to slough off nineteenth-century middle-class ideas and the teachings of Christian philosophy in particular. And since he has made this effort, it is hardly surprising that he should be reluctant to use freely the language of traditional philosophy, let alone of theology. You and I are in quite a different boat—after being involved in two world wars and two revolutions we are able to reject most traditions without very much effort. I should consider it utterly absurd—and Niels, for one, would agree—were I to close my mind to the problems and ideas of earlier philosophers simply because they cannot be expressed in a more precise language. True, I often have great difficulty in grasping what these ideas are meant to convey, but when that happens, I always try to translate them into modern terminology and to discover whether they throw up fresh answers. But I have no principled objections to the re-examination of old questions, much as I have no objections to using the language of any of the old religions. We know that religions speak in images and parables and that these can never fully correspond to the meanings they are trying to express. But I believe that, in the final analysis, all the old religions try to express the same contents, the same relations, and all of these hinge around questions about values. The positivists may be right in thinking that it is difficult nowadays to assign a meaning to such parables. Nevertheless, we ought to make every effort to grasp their meaning, since it quite obviously refers to a crucial aspect of reality; or perhaps we ought to try putting it into modern language, if it can no longer be contained in the old."

"If you think about such problems in that way, then, quite obviously, you cannot accept the equation of truth and predictive power. But what is your own criterion of truth in science? At Bohr's you hinted that it was somehow like the flight of an airplane, but I can't see precisely in what respect. What part of nature is supposed to be analogous to the pilot's intentions or orders?"

"Such terms as 'intentions' or 'orders,' " I tried to explain, "apply to human behavior and can at most serve as metaphors when applied to nature. We may find it more helpful to revert to our old comparison between Ptolemy's astronomy and Newton's conception of planetary motions. If predictive power were indeed the only criterion of truth, Ptolemy's astronomy would be no worse than Newton's. But if we compare Newton and Ptolemy in retrospect, we gain the clear impression that Newton's equations express the paths of the planets much more fully and correctly than Ptolemy's did, that Newton, so to speak, described the plan of nature's own construction. Or to take an example from modern physics: when we learn that the principles of conservation of energy, charge, etc., have a quite universal character, that they apply in all branches of physics and that they result from the symmetry inherent in the fundamental laws, then we are tempted to say that symmetry is a decisive element in the plan on which nature has been created. In saying this I am fully aware that the words 'plan' and 'created' are once again taken from the realm of human experience and that they are metaphors at best. But it is quite easy to see that everyday language must necessarily fall short here. I suppose that is all I can say about my own conception of scientific truth."

"Quite so, but positivists will object that you are making obscure and meaningless noises, whereas they themselves are models of analytic clarity. But where must we seek for the truth, in obscurity or in clarity? Niels has quoted Schiller's 'Truth dwells in the deeps.' Are there such deeps and is there any truth? And may these deeps perhaps hold the meaning of life and death?"

A few hundred yards away, a large liner was gliding past, and its bright lights looked quite fabulous and unreal in the bright blue dusk. For a few moments, I speculated about the human destinies being played out behind the lit-up cabin windows, and suddenly Wolfgang's questions got mixed up with it all. What precisely was this steamer? Was it a mass of iron with a central power station and electric lights? Was it the expression of human intentions, a form resulting from interhuman relations? Or was it a consequence of biological laws, exerting their formative powers not merely on protein molecules but also on steel and electric currents? Did the word "intention" reflect the existence merely of these formative powers or of these biological laws in the human consciousness? And what did the word "merely" mean in this context?

My silent soliloquy now turned to more general questions. Was it utterly absurd to seek behind the ordering structures of this world a "consciousness" whose "intentions" were these very structures? Of course, even to put this question was an anthropomorphic lapse, since the word "consciousness" was, after all, based purely on human experience, and ought therefore to be restricted to the human realm. But in that case we would also be wrong to speak of animal consciousness, when we have a strong feeling that we can do so significantly. We sense that the meaning of "consciousness" becomes wider and at the same time vaguer if we try to apply it outside the human realm.

The positivists have a simple solution: the world must be divided into that which we can say clearly and the rest, which we had better pass over in silence. But can anyone conceive of a more pointless philosophy, seeing that what we can say clearly amounts to next to nothing? If we omitted all that is unclear, we would probably be left with completely uninteresting and trivial tautologies.

My train of thought was interrupted by Wolfgang:

"You said earlier that you were not a stranger to the images and parables of religion and that, for that very reason, you were unhappy about positivist restrictions. You also hinted that, despite differences in their images, the old religions refer to much the same facts, which, as you put it, are intimately bound up with questions of value. Just what did you mean by that, and what precisely have the facts of religion to do with your concept of truth?"

"The problem of values is nothing but the problem of our acts, goals and morals. It concerns the compass by which we must steer our ship if we are to set a true course through life. The compass itself has been given different names by various religions and philosophies: happiness, the will of God, the meaning of life—to mention just a few. The differences in the names reflect profound differences in the awareness of different human groups. I have no wish to belittle these differences, but I have the clear impression that all such formulations try to express man's relatedness to a central order. Of course, we all know that our own reality depends on the structure of our consciousness; we can objectify no more than a small part of our world. But even when we try to probe into the subjective realm, we cannot ignore the central order or look upon the forms peopling this realm as mere phantoms or accidents. Admittedly, the subjective realm of an individual, no less than a nation, may sometimes be in a state of confusion. Demons can be let loose and do a great deal of mischief, or, to put it more scientifically, partial orders that have split away from the central order, or do not fit into it, may have taken over. But in the final analysis, the central order, or the 'one' as it used to be called and with which we commune in the language of religion, must win out. And when people search for values, they are probably searching for the kind of actions that are in harmony with the central order, and as such are free of the confusions springing from divided, partial orders. The power of the 'one' may be gathered from the very fact that we think of the orderly as the good, and of the confused and chaotic as the bad. The sight of a town destroyed by an atom bomb depresses our spirits, a desert transformed into a blossoming meadow refreshes us. In science, the central order can be recognized by the fact that we can use such metaphors as 'Nature has been made according to this plan.' It is in this context that my idea of truth impinges on the reality of religious experience. I feel that this link has become much more obvious since we have understood quantum theory. For quantum theory helps us to formulate orderly processes in a wide field by means of an abstract, mathematical language. And if we try to express these orderly processes in everyday terms, we have to fall back on parables, on complementary viewpoints involving paradoxes and apparent contradictions."

"I can follow you most of the way," Wolfgang said, "but just what do you mean when you say that the central order must win out? It either exists or it does not. What sense can there possibly be in saying that it wins?"

"By that I mean something altogether commonplace, for instance, the fact that as each winter passes, the flowers come into blossom in the meadows, and that as each war ends, cities are rebuilt. Time and again destruction makes way for order."

We walked on in silence and had soon reached the northern tip of the Langvlinie, whence we continued along the jetty as far as the small beacon. In the north, we could still see a bright strip of red; in these latitudes the sun does not travel far beneath the horizon. The outlines of the harbor installations stood out sharply, and after we had been standing at the end of the jetty for a while, Wolfgang asked me quite unexpectedly:

"Do you believe in a personal God? I know, of course, how difficult it is to attach a clear meaning to this question, but you can probably appreciate its general purport."

"May I rephrase your question?" I asked. "I myself should prefer the following formulation: Can you, or anyone else, reach the central order of things or events, whose existence seems beyond doubt, as directly as you can reach the soul of another human being? I am using the term 'soul' quite deliberately so as not to be misunderstood. If you put your question like that, I would say yes. And because my own experiences do not matter so much, I might go on to remind you of Pascal's famous text, the one he kept sewn in his jacket. It was headed 'Fire' and began with the words: 'God of Abraham, Isaac and Jacob—not of the philosophers and sages.' I hasten to add that, in this particular form, the text does not apply to me."

"In other words, you think that you can become aware of the

central order with the same intensity as of the soul of another person?"

"Perhaps."

"Why did you use the word 'soul' and not simply speak of another person?"

"Precisely because the word 'soul' refers to the central order, to the inner core of a being whose outer manifestations may be highly diverse and pass our understanding."

"I am not sure whether I am completely with you. After all, we must not exaggerate the importance of our own experiences."

"Certainly not, but science, too, is based on personal experience, or on the experiences of others, reliably reported."

"Perhaps I ought to have phrased my question differently. But let's rather return to our original problem, the positivist philosophy. It does not appeal to you, because it would prevent discussion of any of the subjects we have just broached. But does that mean that positivism is quite divorced from the world of values? That it cannot include ethics on principle?"

"It would look like that at first sight, but, historically speaking, it was probably quite the other way round. After all, modern positivism has developed out of pragmatism, which teaches us not to sit by with folded arms but to take responsibility for our own actions, to lend a hand wherever we can, even if it does not mean changing the world overnight. In this respect, pragmatism strikes me as superior to many of the old religions. For the old doctrines lead one far too easily into passive resignation and persuade one to bow to the inevitable, when, in fact, one could improve a great many things by one's own actions. Learning to walk before you can run is a very good principle in the practical world, and even in science provided only you do not lose sight of the wider reality. Newton's physics, too, was, after all, compounded of the careful study of details and an over-all view of nature. Unfortunately, modern positivism mistakenly shuts its eyes to the wider reality, wants to keep it deliberately in the dark. I may be exaggerating, but, at the very best, positivism does not encourage people to reflect on this subject."

"Your critique is fair enough, but you have failed to answer my question. If this mixture of pragmatism and positivism does involve an ethical doctrine—and you are certainly correct that it does and that we can see it at work in America and England—by what compass does it set its course? You have claimed that in the final analysis our compass must be our relationship with a central order, but where can you find such a relationship in pragmatism?"

"I agree with Max Weber that, ultimately, pragmatism bases its ethics on Calvinism, i.e., on Christianity. If we ask Western man what is good and what is evil, what is worth striving for and what has to be rejected, we shall find time and again that his answers reflect the ethical norms of Christianity even when he has long since lost all touch with Christian images and parables. If the magnetic force that has guided this particular compass— and what else was its source but the central order?—should ever become extinguished, terrible things may happen to mankind, far more terrible even than concentration camps and atom bombs. But we did not set out to look into such dark recesses; let's hope the central realm will light our way again, perhaps in quite unsuspected ways. As far as science is concerned, however, Niels is certainly right to underwrite the demands of pragmatists and positivists for meticulous attention to detail and for semantic clarity. It is only in respect to its taboos that we can object to positivism, for if we may no longer speak or even think about the wider connections, we are without a compass and hence in danger of losing our way."

Despite the late hour, a small boat made fast on the jetty and took us back to Kongens Nytorv, whence it was easy to reach Niels' house.

18

Scientific and Political Disputes (1956–1957)

Ten years after the end of the war the worst damage had been repaired. The work of reconstruction, at least in West Germany, had made so much headway that it became possible to think of German participation in atomic energy projects. In the autumn of 1954, on the orders of the Federal Government, I went to Washington and took part in preliminary talks on the resumption of this type of work in the Federal Republic. The fact that Germany had made no attempts to build atom bombs during the war, although she had not lacked the necessary skills and knowledge, probably had a favorable effect on these negotiations; in any case, we were given permission to build a small atomic reactor, and it looked very much as if all restrictions on peaceful atomic developments in Germany would soon be lifted.

In these circumstances, the Federal Republic had to set fresh guidelines. The first job, of course, was the construction of a research reactor, in which physicists and engineers, and more generally German industry at large, could study the new techniques. It seemed obvious that our particular division of the Göttingen Max Planck Institute for Physics and Astrophysics, led by Karl Wirtz, would play an important role in this project, not only because it was familiar with the wartime reactor program, but also because its members had avidly followed subsequent developments in the literature or at scientific meetings. For that very reason, Adenauer quite often invited me to be present at official discussions or at government negotiations with German industry, so as to ensure that only scientifically sound plans would be adopted. For me, it was a completely new, though not astonishing, experience to discover that even in a democracy such important decisions as the opening of an atomic branch of industry cannot be based on efficiency alone, but call for the delicate balancing of different interests, which are difficult to discern and often prevent the adoption of the most effective solution. It would be unfair to blame the politicians for this state of affairs; on the contrary, blending conflicting interests into a harmonious and vital social whole is one of their most important and praiseworthy tasks. However, I myself had no experience in balancing economic or political interests, and I could therefore contribute far less to the negotiations than I might have hoped.

From conversations with my closest collaborators, I had gained the impression that it would be a good idea to build the first research reactor in the immediate vicinity of our Institute. This meant that the Institute itself would have to find a much larger site, and I was in favor of a move to the environs of Munich. Admittedly, I was so for personal motives as well—I had the happiest memories of the city from my early youth and student days. But quite apart from that, the proximity of so important a cultural center, one so open to new ideas, as Munich, struck me as being eminently useful to the Institute. As for the relationship between the Institute and the new-born Center of Atomic Technology, I felt that close collaboration was the best way of ensuring that the experiences the Institute had gathered during the war would be put to the best use, and that the atomic engineers trained by our Institute would concentrate on peaceful developments and not be tempted to use the considerable resources of the Center for other purposes. It did not, however, take me very long to realize that the most influential industrialists in our country did not wish to see such important work being done in Bavaria; rightly or wrongly, they thought that Württemberg was much more suitable, and their final choice accordingly fell on Karlsruhe. At the same time, the Bavarian authorities offered a new building to the Institute itself in Munich, but without Karl Wirtz, who had been asked to take his reactor team to Karlsruhe. Carl Friedrich was appointed professor of philosophy at the University of Hamburg.

I was not very happy about these developments, which, though they met my personal wishes with respect to the Bavarian move, completely ignored the advantages of practical atomic research in the proximity of our Institute. I was sad to see my close and long collaboration with Carl Friedrich and Karl Wirtz coming to an end, and I was worried that the new Karlsruhe Center might, sooner or later, fall into the hands of people not primarily concerned with the peaceful uses of atomic energy. I was also disturbed to find that for those who had to make the most important decisions the boundaries between peaceful and military applications, no less than between applied and fundamental research, were extremely fluid.

These anxieties were increased by the fact that politicians and economists, unlike the man in the street, would occasionally express the view that atomic weapons had become an accepted means of national defense and as such had their part to play in the Federal Republic just as they did in the rest of the world. I, like most of my friends, on the other hand, was convinced that atomic weapons would only weaken the international standing of the Federal Republic, that by clamoring for atomic bombs we should simply be damaging our cause. For the outside world still remembered the war atrocities with horror, and public opinion abroad would never have tolerated a Germany brandishing atom bombs. I was therefore pleased to gather from the several talks I had with the Federal Chancellor, that he, too, believed that the Federal Republic should never do more in the field of rearmament than her allies demanded of her. But in this area as well the government had to try its best to balance what were often widely opposed interests.

Of all my friends, Carl Friedrich was the one who returned to this theme most consistently, and who eventually felt compelled to step into the political limelight. One of our many conversations must have started when I asked him: "What do you really think of the future of our Institute? I am very worried about the decision to take the atomic energy project completely out of our hands. Of course, I realize that there are plenty of other tasks for us to tackle, but who wants this separation? Was it perhaps my rather selfish proposal of a Munich site that was the root of all the trouble? Or are there more factual reasons why the planned Atomic Center should be cut off from the Max Planck Society?"

"In all such semipolitical questions," Carl Friedrich said, "it is extremely difficult to tell what is factual and what is not. The technical developments we foresee are bound to cause far-reaching economic changes in the particular area selected for the new plant. Many people will find employment there, and quite possibly new housing developments will have to be built, and subsidiary industries will spring up and get their order books filled. Hence there are good 'factual' reasons why a particular city or region should wish to be chosen for such developments. No doubt, you still remember our discussion at Farm Hall on the subject of the atom bomb. What we said there applies here as well: the choice of a site must fit into the over-all scientific and economic development plans of the Federal Republic. It is not enough to ask where the reactors can be built most efficiently. Other factors ensuring the smooth running of the whole economy have to be taken into consideration as well."

"True enough. And do you think such considerations were, in fact, paramount when Germany's atomic power projects were drawn up?" I asked.

"I don't honestly know, and that is where my real worries begin. As you no doubt have realized from our many talks, it is extremely difficult for most outsiders to draw a sharp dividing line between development plans and fundamental research or weapon technology. Hence there will be attempts—though this need not perhaps worry us too much—to open the Center to basic research with no direct bearing on current technical developments. Again—and this would be much worse—there might be attempts to encourage peaceful atomic projects for the sake of their possible military applications, for instance, in connection with the production of plutonium. Karl Wirtz will certainly do his utmost to halt this trend, but there might well be powerful forces pulling in the opposite direction, so powerful, in fact, that no individual can stop them. Hence we ought to do everything in our power to obtain a binding declaration from the government that there will be no attempt to produce atomic weapons in Germany. Needless to say, the government will want to keep all their options open, and are most unlikely to let us tie them down. We might issue a public declaration, but people may have lost faith in such pronouncements. But then you and a number of physicists got together on Mainau Island last year and signed a joint declaration. Were you satisfied with the results?"

"I did sign the declaration, but I really hate that sort of thing.

To state publicly that you are for peace and against the atom bomb is, after all, nothing but silly chatter. Every human being in his right mind must obviously be for peace and against the bomb, and he does not need us scientists to tell him so. The government will simply include all such protests in their political calculations. They will declare themselves wholly in favor of peace and against the atom bomb, and simply add the subsidiary clause that, of course, they mean the kind of peace that is favorable to, and honorable for, our own people, and that they are, meanwhile, doing their utmost to defend us from the threat of foreign atom bombs. And so we have not moved a single step forward."

"Nevertheless, the people will be reminded of the horrors of an atom war, and if that reminder had been pointless, you would hardly have signed the Mainau declaration."

"Perhaps; still, the more general and less binding a declaration, the less valuable it is."

"Very well, we shall just have to think of something better, but we simply have to do something."

"Most people continue to look on the old politics—the expansion of economic and political power, blackmail through threats of armed force—as the most realistic approach, and this despite the fact that it has long since become the very converse. Quite recently I heard a member of our Federal Government say that if France has atomic weapons, then the Federal Republic has the right to ask for them, too. Needless to say, I contradicted him at once. But the terrifying thing about this argument was not so much his demand as his basic premise: he took it for granted that the possession of atomic weapons is a great political advantage and that the only problem is how to get them. He would have looked on anyone who thought otherwise or even questioned his premises as a hopeless dreamer—or a swindler hiding his own political aspirations, for instance a desire to see the Federal Republic absorbed by Russia."

"You exaggerate because the man annoyed you. The policy of our government is certainly much more reasonable than that, and, when all is said and done, there are many intermediate stages between having an atomic bomb of our own and complete reliance on others. Still, we must do everything we can to prevent things from moving in the wrong direction."

"That will be very difficult. If I have learned anything from developments during the last few months, it is this: you cannot be a good politician and a good scientist at the same time. Nor could it be otherwise. What matters in both is wholeheartedness; anything short of that does no good at all. Hence I shall probably devote myself exclusively to science once again."

"If you do, you will be doing the wrong thing. Politics is no longer the preserve of specialists, and if we want to prevent a repetition of the 1933 disaster, every one of us must play his part. We can't just abdicate; certainly not when what is at stake is the development of atomic physics."

"Very well then. If you need me, you have only to say so."

In the summer of 1956, when this conversation took place, I felt tired and utterly exhausted. Among other things, I was depressed by an argument with Wolfgang Pauli, who refused to share my views on an important scientific problem. During a meeting in Pisa a year earlier, I had made rather unconventional suggestions regarding the possible mathematical formulation of a theory of elementary particles, which Wolfgang could not accept. He had even investigated the closely related mathematical model constructed by the Chinese-American physicist, Lee, and had reached the conclusion that I was barking up the wrong tree. I refused to believe that, and Wolfgang criticized me with all the severity he reserved for such occasions.

"These comments," he wrote to me from Zurich, "are mainly intended to demonstrate to you that at the time of the Pisa conference you understood next to nothing about your own work."

At first, I was too exhausted to do justice to the mathematical problems involved, and I accordingly decided on a fairly long rest. To that end, I retired with my whole family to Liselje, the small seaside resort on the island of Zealand, only some six miles from Bohr's summer home in Tisvilde. I wanted to use the opportunity to spend a great deal of time with Niels, without having to make too severe demands on his hospitality. These were happy weeks indeed. Mutual visits helped to cure my fatigue and gave me a chance to restore the links with our common past. As for my mathematical controversy with Wolfgang, Niels, understandably, did not wish to become involved—he did not feel competent to pronounce on questions that were more of a mathematical than of a physical nature. However, he was in agreement with the general philosophical premises on which I wanted to base the theory of elementary particles, and he encouraged me to proceed in the direction I had set out on.

A few weeks after my return from Denmark, I fell seriously ill and had to keep to my bed for quite some time. Working was out of the question, nor could I be present at the political discussions when Carl Friedrich and other friends hammered out our demands to the government. The day I got up—it was the end of November by then—the "18 Göttingers," as we were later called, met in my house and drafted a letter to the then Minister of Atomic Energy, Franz-Josef Strauss. In it we said that, unless we received a satisfactory answer to our letter, we reserved the right to put our views on atomic rearmament before the public. I was glad now that Carl Friedrich had taken the initiative in this matter, for all I myself was fit to do at the time was to watch and give him my moral support.

In the following weeks, in which I recovered very slowly, I tried to resolve my controversy with Wolfgang. Basically, it hinged on my proposal to extend, for the formulation of the physical laws governing the behavior of elementary particles, the mathematical space that had served for this purpose ever since the introduction of quantum mechanics, and to which physicists referred somewhat loosely as the Hilbert space. The suggestion that this space be extended by admitting a so-called "indefinite metric" had come from Paul Dirac thirteen years earlier, but Wolfgang had then proved that the magnitudes which quantum mechanics interprets as probabilities can, in this case, occasionally assume negative values and that the mathematical result would no longer make physical sense. At about the time of the Pisa conference, he had gone on to produce detailed mathematical objections to this attempt, using the model Lee had proposed. I, for my part, had tried to resurrect Dirac's suggestion, claiming that, in special cases, which I described, Wolfgang's objections could be met. Wolfgang naturally refused to believe that this was so.

I now decided to apply Wolfgang's own mathematical method, namely, his analysis of Lee's model, in an attempt to demonstrate that the difficulties could be avoided in the special case I had mentioned. It took me until the end of January before I was ready to produce my proof in a letter to Wolfgang. At the same time, unfortunately, my health deteriorated once again, to the extent that the doctor advised me to leave Göttingen and to let Elisabeth look after me in Ascona on Lake Maggiore until I had fully recovered. My correspondence with Wolfgang from Ascona remains a most painful memory. Both of us fought remorselessly and summoned up all our mathematical resources in an attempt to break the deadlock. At first, my proof was not yet fully clear and Wolfgang could not see what I was getting at. Again and again, I tried to re-present my arguments, and each time Wolfgang was incensed at my failure to see his objections. Finally his patience was almost exhausted, and he wrote: "That was a terrible letter you sent me. Nearly everything in it strikes me as hopelessly mistaken. . . . You keep repeating your fixed ideas or false conclusion, as if you never bothered to read what I write to you. It seems I have just been wasting my time, and the best thing I can do now is to put an end to this futile discussion." But I refused to leave matters there, and though my illness kept recurring, with serious attacks of vertigo and depression, I was determined to make my point.

After nearly six weeks of the most intense effort, I finally succeeded in breaching Wolfgang's defenses. He now realized that, far from trying to produce a general solution of the mathematical problems under discussion, I was merely offering a series of special solutions for which I claimed only that they lent themselves to a physical interpretation. We had taken the first step toward a reconciliation, and after working through the various mathematical details, both of us were finally satisfied that the problem had been solved, or rather that the unconventional mathematical schema on which I wanted to base the theory of elementary particles did not contain any obvious self-contradictions. Admittedly, this in itself was no proof that my scheme was a useful one, but there were additional reasons for believing that the solution had to be sought along the lines I had followed, and that I was justified in continuing along them. On my return from Ascona, I had a thorough medical check-up at the University Clinic in Zurich, and used the opportunity to fit in a meeting with Wolfgang, which passed very amicably, so much so, in fact, that Wolfgang concluded we had arrived at a "boring unanimity." The Battle of Ascona, as he was to call it in our subsequent correspondence, was definitely over.

I spent the next few weeks in our Urfeld home, where I made a quick recovery. On my return to Göttingen, I learned that political arguments over the atomic rearmament problem were moving toward a crisis point. The authorities refused to take the path we had proposed, and while we appreciated their reasons, we became increasingly anxious lest they move in the opposite direction. Imagine our consternation when Adenauer declared in public that the acquisition of atomic weapons was nothing more nor less than a strengthening of our artillery, and that there was only a difference of degree between nuclear bombs and conventional arms. To our mind this was nothing short of deception, and we felt compelled to do something about it. Carl Friedrich suggested that, as a first step, we make our views public.

All of us quickly agreed that we must not issue just another well-meaning protest against the atom bomb or a vague declaration in favor of peace, but that we should aim at definite objectives, capable of being implemented under the current circumstances. Two of these were quite obvious to us. To begin with, the German people had to be fully informed about the full effects of atomic weapons—they must be spared none of the horrors. Second, we would have to force the Federal Government to change its attitude toward atomic warfare, on the grounds that ownership of atomic weapons, far from adding to our security, put the very survival of the Federal Republic in jeopardy. Our appeal was therefore specifically to the Germans—what other governments or people thought about atomic weapons must be a matter of complete indifference to us in this particular context. Finally, we believed that we could lend weight to our declaration if we, as individuals, solemnly refused every form of participation in any atomic rearmament program. We felt even more entitled to take this stand since we had succeeded in avoiding similar commitments even during the war—admittedly, luck had been on our side. The others worked out the details—I was still under treatment and so was excused from attending most of the meetings. The text of the declaration was drafted by Carl Friedrich and, after modifications worked out during several meetings, approved by all the eighteen Göttingen physicists.

The manifesto was published in the press on April 16, 1957, and obviously had a strong effect on the German public. Our first objective was apparently attained after only a few days—no one tried to suggest that we had exaggerated the effects of atomic weapons. The Federal Government itself seemed divided. Adenauer was upset by a campaign that ran counter to the course he had carefully chosen, and he invited several of us, myself included, to a discussion in Bonn. I refused, because I could not imagine that any fresh or helpful ideas would emerge, and also because, for health reasons, I felt that I ought to avoid strenuous arguments. Adenauer telephoned and begged me to change my mind, and we had a long political discussion, roughly along the following lines:

Adenauer began by pointing out that we had always agreed on all questions of principle, that peaceful atomic developments were proceeding apace in the Federal Republic, and that our appeal was probably based on a series of misunderstandings. Under the circumstances he had a right to expect that we pay careful attention to his arguments, to the reason why he was demanding greater elbow room with respect to atomic weapons. He was certain that, once we had heard these arguments, we would quickly see his point, and it was very important to him that this fact be given as much publicity as our original manifesto. I replied that I had been ill and that I did not yet feel strong enough to argue with him on such critical questions as atomic rearmament. Nor did I believe that agreement could be reached as quickly as he assumed, since all he was likely to tell us was that the Federal Republic was suffering from military weakness vis-à-vis the Soviet Union, and that it was immoral to expect the Americans to defend us if we ourselves were not prepared to make considerable sacrifices. Now we had thought very deeply about all these arguments, and we were possibly in a better position than most of our compatriots to judge how people in America and England felt about us Germans. My travels during the past few years had left me in no doubt that any attempt to equip the German Army with atomic weapons would produce a loud stream of protest, particularly in America, and that the resulting drop of the political temperature, which was cold enough as it was, would more than outweigh the possible military advantage.

Adenauer replied that he knew we physicists were idealists who believed in man's inherent goodness and detested the use of force. He would have been in complete agreement with us had we issued a world-wide appeal against atomic weapons and a general call for the peaceful settlement of conflicts. These were precisely his own aims. However, in addressing ourselves specifically to the German people, we had made it look as if we had set out deliberately to weaken the Federal Republic. In any case, this might well be the effect of our manifesto.

I took the stongest exception to this line. I was certain, I said, that in this of all matters we had acted as sober realists, not as idealists. We were convinced that the use of atomic weapons by the Federal Army would dangerously weaken Germany's political standing in the world; that our security, with which he was so rightly concerned, would be most gravely endangered by atomic rearmament. We were living in an age in which defense problems changed as radically as, for instance, they had during the waning of the Middle Ages, and all of us ought to reflect deeply about these changes before blithely reverting to the old thought patterns. Our manifesto had no other aim than to make people think along these lines and to prevent outmoded tactical considerations from leading us astray once again.

Adenauer found it difficult to follow my arguments, and thought it improper that a small group of people, in this case atomic physicists, should presume to interfere in plans he had so carefully laid to accord with the interests of large political groups. At the same time, he must have realized from the public response to our manifesto that a considerable section of the German people was on our side—no less than people in other parts of the world—and that our manifesto could not simply be swept under the rug. He tried once again to persuade me to come to Bonn, but finally realized that he might be pressing me too hard.

I cannot tell how annoyed with us Adenauer really was at the time. A few years later he sent me a letter in which he said that he was quite capable of respecting political opinions differing from his own. But he was probably a skeptic at heart, and hence quite clear about the narrow limit that is set to all political actions. It pleased him, moreover, to look for passable roads within these limits, and he was disappointed whenever these proved rougher than he had imagined. His own compass was not set by the Prussian model of which I had spoken to Niels so many years ago during our hike through Denmark, nor did he share the libertarian views of the Icelandic sagas so dear to the British. Adenauer's guideline was the Roman-Christian tradition as it had survived in the Catholic Church, together with certain nineteenth-century social doctrines whose Christian roots he seemed to recognize despite the apparent taint of Communism and atheism. The Catholic faith contains a good dose of Eastern wisdom, and it was precisely this wisdom from which Adenauer drew strength in times of difficulty. I remember a conversation about our several experiences in P.O.W. camps. Adenauer had been incarcerated in a Gestapo prison with starvation rations, while I myself had had a relatively pleasant time in England, and so I asked him one day whether he had suffered a great deal. "Well," he said, "when you are locked up in a small cell for days, weeks, months, and are never disturbed by telephone calls and visitors, you can think back at leisure on the past and reflect quietly on what may still be in store for you, and that is really quite a nice thing to be able to do."

19

The Unified
Field Theory (1957–1958)

In the harbor of Venice, opposite the Doge's Palace and the Piazzetta, lies the Island of San Giorgio. It is owned by Count Cini, who runs a school in which orphans and foundlings are trained as sailors and craftsmen, and who has restored the famous old Benedictine monastery on the island. A few glorious rooms on the first floor of that monastery have been set aside as guest rooms, and during the Atomic Physics Congress in Padua in the autumn of 1957 the Count was kind enough to invite some of the older delegates, including Wolfgang and myself, to stay on his island. The quiet cloisters, in which the noise of the busy port could only just be heard, and occasional drives to Padua gave us many a good opportunity for conversations about current developments in atomic physics.

Chief among these was the work of the young Chinese-American physicists Lee and Yang. These two had put forward the suggestion that mirror or right-left symmetry, always considered an almost self-evident part of nature, could be disturbed by such weak interactions as are, for example, responsible for radioactive phenomena. And, in fact, Madame Wu's later experiments demonstrated quite clearly that radioactive beta decay is accompanied by deviations from that symmetry. It looked very much as if the weightless particles emitted during beta decay—the so-called neutrinos—existed in only one form, let us call it the left-hand form, while antineutrinos occurred only in the right-hand form. Wolfgang was particularly interested in the properties of neutrinos because he had been the first to predict their existence twenty years earlier. He had long since been proved correct, but Lee's and Yang's discovery called for drastic and exciting modifications of the old model. We, that is, Wolfgang and I, had always thought that the symmetries represented by these simple weightless particles must also be properties of the underlying natural law. Now, if these particles were indeed devoid of mirror symmetry, then we had to reckon with the possibility that the latter was not a primary aspect of the fundamental laws of nature but entered them secondarily by way of, for example, interactions and the resulting mass. In that case, mirror symmetry would originate in a subsequent doubling-up process that could arise mathematically, for instance, through the fact that an equation admits of two equivalent solutions. This possibility seemed very exciting to us, simply because it amounted to a simplification of the fundamental laws of nature. We had long ago learned from our studies that whenever an unexpected simplification appears during an experiment, we must pay careful attention to it: for here we may have reached a point from which we can catch a glimpse of the wider connections. In the present case, we had the feeling that the Lee-Yang discovery might easily lead us to decisive new insights.

Lee himself happened to be at the Congress and seemed to share this view: during a long conversation in our cloisters, he told me that he, too, was hopeful that important new connections were "just around the corner." Wolfgang was very confident as well, partly because he was so familiar with the mathematical structures associated with neutrinos and partly because our earlier discussions during the "Battle of Ascona" had made him optimistic about the construction of self-consistent relativistic quantum field theories. He was particularly fascinated by the process of doubling-up or division, which, he believed, was responsible for the appearance of mirror symmetry—even though it was still impossible to express this fact in concrete mathematical terms. It was thanks to this process that nature—in a way still to be investigated—introduced a new symmetry property. As for the subsequent disturbance of the symmetry, we were even vaguer about it than about the division. Nevertheless, we would occasionally express the view that the universe as a whole did not necessarily have to be symmetrical with respect to the operations under which natural laws remain invariant; in other words, that the observed reduction in symmetry might possibly be the result of a cosmic asymmetry. At the time, all such ideas were very much less clear in our heads than they may look in writing. In any case, they exerted a great fascination on us all, and once we had begun to think about them, it became quite impossible to shake them off. One day, when I asked Wolfgang why he laid so much stress on the doubling process, he made the following reply:

"In the earlier physics of the atomic shell we had to rely exclusively on perceptual models taken from the arsenal of classical physics. Bohr's correspondence principle stressed the usefulness, however limited, of such models. But the mathematical description of what goes on in the atomic shell was always much more abstract than such models. In fact, it is quite possible to attribute quite different and mutually contradictory models, for example, the particle and the wave models, to the same real process. In the physics of elementary particles, however, all such models prove of no practical use at all, for that branch of science is even more abstract. If we wish to formulate the physical laws in this realm, we must therefore base ourselves on the properties of symmetry that nature herself has introduced here, or, to put it differently, on the symmetry operations (for instance, displacements and rotations) that open up nature's space. Now this forces us to ask why there are these symmetry operations and no others. I think that the concept of division or doubling will prove particularly useful here, because it helps to extend nature in what seems to be an unforced manner, and may thus introduce new symmetries. In the ideal case, we could imagine that all real symmetries have come about as a result of this kind of division."

Real work on these problems had, of course, to be deferred until after our return from the Congress. Back in Göttingen I concentrated my own efforts on a search for a field equation governing the behavior of a material field with internal interactions and, if possible, representing all the symmetries that can be observed in nature. As a model I used the interaction characteristic of beta decay, which had received its simplest, and probably definitive, form through the discovery of Lee and Yang.

In the late autumn of 1957, I had to give a lecture on these and related problems in Geneva, and on my way back I stopped briefly in Zurich for a discussion with Wolfgang, who encouraged me to proceed in the direction I had taken. This was just what I needed, and in the next few weeks I kept examining a host of different forms in which the internal interactions of the material field could be represented. Quite suddenly, there appeared among the fluctuating forms a field equation with a very high degree of symmetry, and hardly more complicated than Dirac's old electron equation. However, besides the space-time structure of relativity theory, it also contained the proton-neutron symmetry of which I had first dreamed on our skiing holiday during Easter 1933. Or, to put it in more mathematical terms, it contained not only the Lorentz group but also the isospin group—in other words, it seemed to account for a great many symmetries found in nature. Wolfgang, whom I informed of the latest development, was extremely excited as well: it really did look as it, for the first time, we had a framework wide enough to include the entire spectrum of elementary particles and their interactions, and yet narrow enough to determine everything in this field apart from contingent factors. And so we decided that both of us would look into the question of whether or not this equation might serve as a basis for a unified field theory of elementary particles. Wolfgang was hopeful that what few symmetries were still missing might be added later by means of the division process.

With every step Wolfgang took in this direction, he became more enthusiastic—never before or afterward have I seen him so excited about physics. And much as, in past years, he had been critical and skeptical of all our theoretical efforts, which had admittedly borne on no more than partial aspects of the physics of elementary particles and not on their over-all connection, so he was now determined to express that very connection by means of the new field equation. He became firmly convinced that our equation, unique in its simplicity and high symmetry content, must be the right starting point for the unified field theory of elementary particles. I, too, was fascinated by the new possibilities, which looked like the golden key to the gate that had hitherto barred access to the world of elementary particles. Just the same, I was only too keenly aware of the many difficulties in our path. Shortly before Christmas 1957, I received a letter from Wolfgang. It reflected his great elation during those weeks:

. . . Division and reduction of symmetry, this then the kernel of the brute! The former is an ancient attribute of the devil (they tell me that the original meaning of *"Zweifel"* [doubt] was *"Zweiteilung"* [dichotomy]). A bishop in a play by Bernard Shaw says: "A fair play for the devil, please." So let him join us for Christmas. If only the two divine contenders—Christ and the devil—could notice that they had grown so much more symmetrical! Please don't repeat this heresy to your children, but you can mention it to Baron von Weizsäcker.

Very, very cordially yours,
WOLFGANG PAULI

And a week later he scribbled over his salutation: "All the best to you and your family for the New Year. Let's hope it will bring us complete understanding of the physics of elementary particles." And in the letter itself, he said:

The picture keeps shifting all the time. Everything is in flux. Nothing for publication yet, but it's all bound to turn out magnificently. No one can tell just what marvels will appear. Wish me luck, I am learning to walk. [And then the quotation:] Reason begins again to speak, again the bloom of hope returns. The streams of life we fain would seek, ah, for life's source our spirit yearns. Greet the dawn of 1958 before sunrise. . . . Enough for today. This is powerful stuff. . . . The cat is out of the bag, and has shown its claws: division and symmetry reduction. I have gone out to meet it with my antisymmetry—I gave it fair play—whereupon it made its quietus. . . . A very happy New Year. Let us march toward it. It's a long way to Tipperary, it's a long way to go.

Cordially yours,
WOLFGANG PAULI

Needless to say, Wolfgang's letters also contained a great many mathematical details, but this is hardly the place to discuss them.

A few weeks later, Wolfgang was due to leave for America, where he had lecture engagements for three months. I did not like the idea of this encounter between Wolfgang in his present mood of great exaltation and the sober American pragmatists, and tried to stop him from going. Unfortunately, his plans could no longer be changed. We then prepared a draft for a joint publication, and, as is customary, sent it to several interested friends. Then we were divided by the Atlantic, and Wolfgang's letters came at greater and greater intervals. I thought I noticed signs of fatigue and resignation in them, but otherwise Wolfgang was apparently still determined to see our common project through. Then, quite suddenly, he wrote me a somewhat brusque letter in which he informed me of his decision to withdraw from both the work and the publication. He added that he had informed the recipients of the preliminary draft that its contents no longer represented his present opinion. He gave me full authority to do what I liked with the results. Then the correspondence was broken off, and I failed to get any further information from Wolfgang about his sudden change of mind, though I assumed that he had become discouraged by the lack of clarity of the thought processes involved. But this did not fully explain his behavior. I myself was only too aware of the difficulties, but we had often worked together in the dark, and as far as I myself was concerned such situations had always struck me as the most interesting.

I did not meet Wolfgang again until July 1958, when both of us attended a congress in Geneva. I was due to give a report on the current state of research into the disputed field equation, and Wolfgang's attitude to me was almost hostile. He criticized many details of my analysis, some, I thought, quite unreasonably, and he could barely be persuaded to discuss matters with me at greater length. Later, we had one more, fairly long meeting in Varenna on Lake Como. Here, regular summer schools are held in a villa, whose terraced gardens overlook the center of the lake, and since the subject this time was the physics of elementary particles, Wolfgang and I were among the invited guests. Wolfgang was cordial again, but I felt he was a changed man. We would take long walks on the rose-bordered path separating park and lake, or sit on a bench amid a profusion of flowers and look across the blue water to the peaks of tall mountains. On one such occasion, Wolfgang again referred to our common hopes.

"I think you are doing right to continue working on these problems," he said. "You know how much remains to be done, and things will no doubt come out right one day. Perhaps all our hopes will be fulfilled, and your optimism will be rewarded. As for me, I have had to drop out, I just haven't the strength. Last Christmas I still thought I was fit enough for anything, but I can say that no longer. Let's hope you are still up to it, or else that some of your younger colleagues will carry on the job. You seem to have several excellent young physicists in your Institute. But it's all quite beyond my strength, and that's that."

I tried to console him. I said that he was probably disappointed to have found that he could not run as fast as he had anticipated last Christmas, but that once he got down to the work itself, his old courage would quickly return.

"I'm afraid not. Things have changed too much," was all he said.

On one occasion, Elisabeth, who had accompanied me to Varenna, expressed her deep anxiety about Wolfgang's health. She thought he was terribly ill, but I failed to see it. We never saw Wolfgang again.

Toward the end of 1958 I received the sad news that he had died after a sudden operation. I cannot doubt but that the beginning of his illness coincided with those unhappy days in which he lost hope in the speedy completion of our theory of elementary particles. I do not, of course, presume to judge which was the cause and which the effect.

20

Elementary Particles and
Platonic Philosophy (1961–1965)

When the Max Planck Institute for Physics and Astrophysics, which my colleagues and I had helped to build up in Göttingen after the war, moved to Munich in the autumn of 1958, a fresh chapter was opened in all our lives. A new generation of physicists was able to work on the problems that had engaged so much of my attention in a modern, spacious building situated in the northern part of the city, on the edge of the English Garden, and designed by Sep Ruf, a friend from the Youth Movement. The unified field theory of elementary particles became the special concern of Hans Peter Dürr, who had grown up in Germany, had trained in the United States and who, after working under Edward Teller in California for a long time, had returned to his native land. Teller must have told him about our Leipzig circle, and once back in Munich he acquired a liking for the old traditions through conversations with Carl Friedrich, who spent a few weeks in our Institute every autumn, lest his philosophical studies lead to a complete neglect of physics. As a result, physical and philosophical aspects of the unified field theory were the subject of many discussions among the three of us, generally in my study in the new Institute. I shall record one of these discussions as a typical example.

Carl Friedrich: "Have you made progress with your unified field theory since our meeting last year? I mean with regard to the interpretation of the experiments, though I myself am of course particularly interested in the philosophical aspects. But then a theory like yours is solid physics or it is nothing at all. Well, did you get any further, or, more specifically, have you discovered anything new in connection with Pauli's idea of 'division and symmetry reduction'?"

Dürr: "We think that we have understood division in at least one case, namely, mirror symmetry. It comes about because, in relativity theory, the eigenvalue equation for the mass of an elementary particle must be a quadratic equation and therefore have two solutions. But the reduction of symmetry is more interesting still. It seems that here we come face to face with very general and most important relations that we had failed to take into consideration. If one of nature's fundamental symmetries is regularly found to be disturbed in the spectrum of elementary particles, the only possible explanation is that the universe, i.e., the substratum where the particles originated, is less symmetrical than the underlying physical law. Now that is perfectly possible and can be reconciled with the symmetrical field equation. It follows—I do not wish to prove it now—that there must be forces acting over long distances, or elementary particles of vanishing inertial mass. This is probably the best way of interpreting electrodynamics. Gravitation, too, could arise in this way, so that here we may hope to find a bridge to the principles on which Einstein wanted to base his unified field theory and cosmology."

Carl Friedrich: "If I have understood you correctly, you assume that the form of the universe is not absolutely determined by the field equation. In other words, you believe that the universe could exist in various forms, all of them in accordance with your field equation. That would mean, would it not, that your theory contains a contingent element, i.e., that it involves chance or rather an inexplicable and unique factor? This is not at all astonishing from the viewpoint of the older physics since there, too, the initial conditions are not determined by physical laws but are contingent—that is, they might have been otherwise. A mere glance at the present form of the universe, at the countless galactic systems with their random distribution of stars and stellar systems, suggests almost irresistibly that things might have been quite different, that the number and position of the stars, the number and size of the galactic systems, might have had quite different values, and yet the same physical laws could still apply.

Luckily, when dealing with the spectrum of elementary particles we do not have to bother about details on the cosmic scale. But you believe that the general symmetry properties of the universe do have repercussions on that spectrum. Perhaps we could represent such general properties by simplified models of the universe, as we do in general relativity. The underlying field equation would admit some of these models and bar others, and the spectrum of elementary particles might look slightly different for each one. In that case we could conclude from the spectrum of the elementary particles as to the universal symmetries."

Dürr: "Yes, that is precisely what we hope to do. Quite recently, for instance, we made certain assumptions about these symmetries which were later refuted by new experiments involving certain elementary particles. We accordingly replaced the original assumptions with others that did agree with the experimental results. At present, it looks very much as if we can interpret the whole of electrodynamics in terms of the asymmetry of the universe vis-à-vis the proton-neutron exchange or more generally vis-à-vis the isospin group. Here the unified field theory seems to have enough flexibility to allow the inclusion of all the observed phenomena into a general framework."

Carl Friedrich: "If one keeps thinking along these lines, one arrives at a very interesting and difficult problem. I think that in the case of contingent processes we must make a basic distinction between the unique and the accidental. The universe is unique, so that in the beginning unique decisions were made about its symmetries. Later, when a host of galactic systems and stars were being formed, the same decisions had to be made over and over again, decisions which, in a sense, may be called accidental, precisely because they are so numerous and repeatable. Only to them do the statistical laws of quantum mechanics truly apply. Admittedly, the use of expressions such as 'in the beginning' and 'later' is questionable here, because the concept of time, too, is only given a clear meaning by the actual model of the universe. But perhaps we ought to ignore this objection for the purposes of the present discussion. The unique decisions which, as it were, come at the beginning, also include the laws of nature you are trying to describe with your field equation. For we are entitled to ask why these laws have a particular form and no other; just as we are entitled to ask why the universe has certain symmetrical properties and no others. Perhaps there are no answers to these questions, but I find it rather unsatisfactory to take your field equation for granted, even if it is distinguished from all others by its great symmetry and simplicity. Perhaps your equation may acquire an even deeper significance with the help of Pauli's 'division and symmetry reduction.' "

"I am the last person to deny that," I said. "But for the moment I should like once again to stress the uniqueness of the first decisions. They determine symmetries once and for all, and lay down forms that have a decisive effect on subsequent events. 'In the beginning was symmetry' is certainly a better expression than Democritus' 'In the beginning was the particle.' Elementary particles embody symmetries; they are their simplest representations, and yet they are merely their consequence. Accident came later on in the development of the universe, but it, too, fits neatly into the original forms; it satisfies the statistical laws of quantum theory. In the later, ever more complicated, course of events this process can be repeated: unique decisions can once again establish forms that will determine the subsequent events. This, for instance, is what happened during the origin of life, and I, for one, find the discoveries of modern biology extremely illuminating in this respect. The special geological and climatic conditions prevailing on our planet have led to the emergence of a complicated carbon chemistry, with gigantic molecules in which information can be stored. Nucleic acid has proved a suitable store of information for statements about the structure of living beings. With it, a unique decision was taken, and a form established that determined all subsequent biological processes. In these, however, accident once again played an important role. If some planet in another stellar system had the same climatic and geological conditions as prevail on earth, and if there, too, carbon compounds led to the formation of nucleic-acid chains, it still does not follow that precisely the same living beings would people it as live on earth. But what beings there are will have the same nucleic-acid structure. In saying this I am reminded of Goethe's attempt to derive the whole of botany from a single, primordial plant. That plant was said to be an object, but at the same time to represent the basic plan on which all other plants were constructed. Following Goethe, we could call nucleic acid a primordial living being, for it, too, is an object and at the same time represents a biological blueprint. If we talk like that, we are, of course, right back with Plato's philosophy. Our elementary particles are comparable to the regular bodies of Plato's *Timaeus*. They are the original models, the ideas of matter. Nucleic acid is the idea of the living being. These primitive models determine all subsequent developments. They are representative of the central order. And though accident does play an important part in the subsequent emergence and development of a profusion of structures, it may well be that accident, too, is somehow related to the central order."

Carl Friedrich: "I am not at all happy with your use of the word 'somehow.' Perhaps you might care to elaborate. Do you think that this sort of accident is completely pointless? Does it, so to speak, merely put into practice what quantum laws express statistically? Your remarks suggest that, over and above that, you are thinking of a wider connection, a kind of superstructure that lends meaning to the individual event. Am I right in saying that?"

Dürr: "Any deviation from the frequency of events established by quantum mechanics would make nonsense of our explanations why phenomena should otherwise be governed by quantum laws. Experience suggests that such deviations are quite impossible. But you probably had something quite different in mind; no doubt, you were thinking of events or decisions that are essentially unique, i.e., to which statistical considerations do not apply. Still, the use of the word 'meaning' in your question gives it a somewhat unscientific twist."

There the conversation ended. However, a few days later it had a sequel in which I was essentially a passive participant. Konrad Lorenz, Erich von Holst and their collaborators had begun to make a special study of the behavior of the local fauna at the Max Planck Institute for Behavioral Research, situated beside a lake in the hilly and wooded country between Lake Starnberg and Lake Ammer. They had been talking—and this is the title of one of Lorenz' books—with cattle, birds and fishes. The Institute held regular autumn conferences in which biologists, philosophers, physicists and chemists discussed various basic and, above all, epistemological problems of biology. Wits called it the body-soul colloquium. I would occasionally be present, mostly as a listener because I knew very little about biology and was anxious to learn more. I remember that, on the day in question, the talk came around to Darwin's theory in its modern form, i.e., to mutation and selection, and that the speaker contended that the origin of species was comparable to the origin of human tools. Thus man's need to move across the water had led him to invent the rowing boat, and suddenly lakes and coastal waters had begun to teem with new objects. Then someone had the idea of exploiting the force of the wind by means of sails, and sailing boats began to oust rowing boats. Finally, the steam engine was invented and steamships displaced sailing boats on all the oceans. The speaker went on to argue that false starts were quickly suppressed in the course of technical progress. Thus in the history of the light bulb, the Nernst lamp was replaced by the incandescent lamp almost as soon as it had appeared. The process of biological selection must be envisaged in much the same way. Mutations occur by pure chance, just as quantum theory would expect them to do, and selection then eliminates most of these "natural experiments." Only a few forms, which have proved themselves under the given circumstances, remain.

While thinking about this comparison, it occurred to me that the process of technological advance differs from Darwinian theory in one crucial respect, namely, just where Darwinian theory introduces chance. Human inventions are the result never of accident but of man's intention and thought. I tried to see what would happen if the comparison were taken more seriously than the speaker would have wished, and if something like intention were associated with Darwinian mutation. But can one really speak of intentions apart from man? At most, we may be prepared to grant that a dog jumping onto the kitchen table "intends" to eat up the sausage. But has a bacteriophage approaching a bacterium the intention of entering it and of multiplying inside? And even if we are still prepared to say yes, can we also say that genes change their structure with the intention of adapting to their environment? If we did, we would obviously be misusing the word "intention." But perhaps we could choose a more careful formulation. We could ask whether the aim to be reached, the possibility to be realized, may not influence the course of events. If we do that, we are almost back with quantum theory. For the wave function represents a possibility and not an actual event. In other words, the kind of accident which plays so important a role in Darwinian theory may be something very much subtler than we think, and this precisely because it agrees with the laws of quantum mechanics.

My chain of thought was interrupted by a clash of opinion on the platform as to the relevance of quantum theory in biology. Such arguments generally result from the fact that, whereas most biologists are fully prepared to admit that the existence of atoms and molecules can only be understood through quantum theory, they nevertheless tend to look upon atoms and molecules as objects of classical physics—that is, they treat them as if they were so many bricks or grains of sand. This approach may quite often lead to the correct results; but then the conceptual structure of quantum theory differs radically from that of classical physics. Hence thinking along the old lines may occasionally prove misleading. Still, I shall say no more about this part of the "body-soul colloquium."

In my own Institute a number of young physicists were now constantly engaged on problems related to the unified field theory. The stormy disputes of the early years had long since given way to calmer thought. The task before us now was to advance the theory step by step and to fit the known phenomena as coherently into it as we possibly could. Experiments with large accelerators in Geneva and Brookhaven had revealed many fresh details in the spectrum of elementary particles, and we were anxious to ascertain whether or not the new results agreed with our predictions. In this way, the unified field theory gradually assumed a tangible physical form, and Carl Friedrich took a growing interest in its philosophic foundations. Pauli's old theme —division and symmetry reduction—was by no means dead and forgotten. The example discussed by Dürr—mirror symmetry— had merely been a special case, one in which the essential traits of the problem may only have been just discernible. Carl Friedrich now made a determined effort to get at the root of the wider problem.

Many of our discussions during these years took place in

Urfeld. Things had grown more peaceful and quiet, and we had more leisure for retiring to our house on Lake Walchen, particularly at weekends or during the holidays. From the terrace we could look across the lake and the mountains, and delight in the profusion of glowing colors that Lovis Corinth had so lovingly captured on canvas forty years earlier. It was rare now that I was reminded of that other, darker picture: Colonel Pash kneeling behind the terrace wall, his machine pistol at the ready; the sound of shooting down on the road, the children in the cellar behind sandbags. Those events now lay far behind us, and we could once again meditate peacefully about the great questions Plato had once asked, questions that had perhaps found their answer in the contemporary physics of elementary particles.

During one of his visits, Carl Friedrich explained the fundamental ideas of his current attempt: "All our thinking about nature must necessarily move in circles or spirals; for we can only understand nature if we think about her, and we can only think because our brain is built in accordance with nature's laws. In principle, therefore, we could start anywhere at all, but our minds are made in such a way that it seems best to start with what is simplest, namely, alternatives: yes or no, to be or not to be, good or evil. Now, as long as we conceive of these alternatives in the way we do in daily life, that is all there is to it. But as we know from quantum theory, an alternative does not simply amount to a yes or no, but also involves other, complementary answers in which the degree of probability of the yes or no is laid down, as well as their mutual interference. As a result, we have a whole continuum of possible answers or, mathematically speaking, a continuous group of linear transformation of two complex variables. This group contains the Lorentz group of relativity theory. If we ask whether or not any one of these possible answers applies, we are, in fact, asking questions about a space comparable to the space-time continuum of the real world. It is along these lines that I am trying to develop the group structure you have captured in your field equation—and with which the world is, in a sense, unfolded—by the superposition of alternatives."

"In other words, you think it important," I interjected, "that Pauli's division is not a dichotomy in the sense of Aristotelian logic, but that it introduces complementarity in a crucial place.

Pauli was thus right to claim that division in Aristotle's sense was an attribute of the devil; that its continuous repetition can lead only to chaos. But then the new and third possibility, which has appeared with complementarity, may bear fruit and, on repetition, lead us into the space of the real world. As you know, the old mystics used to associate the number 3 with the divine principle. Or if you dislike mysticism, you could think of the Hegelian triplet: thesis–antithesis–synthesis. Synthesis need not be a mixture, a mere compromise between thesis and antithesis; it can prove extremely fruitful, but only when thesis and antithesis combine to produce something qualitatively new."

Carl Friedrich was not altogether satisfied: "Yes, these are pleasant enough philosophical musings, but I'm afraid we can't leave it at that. I am hopeful that the new approach will lead us to the real laws of nature. Your field equation, of which no one can as yet say with certainty whether it represents nature correctly, looks as if it might have originated in this philosophy of alternatives. But surely we ought to be able to establish this fact with the necessary mathematical rigor."

"In any case," I said, "you want to construct the elementary particles, and with them the world, from alternatives in the same way as Plato tried to construct his regular bodies, and the world, from triangles. Your alternatives are no more material than the triangles of Plato's *Timaeus*. But if we start with the logic of quantum theory, then alternatives become the basic form from which more complicated forms are created by repetition. If I understand you, the path leads from alternatives to symmetry groups, that is, to properties. One or several properties are represented by the mathematical forms symbolizing elementary particles; these forms are, so to speak, the ideas of the elementary particles on which the actual particles are modeled. I can follow all this perfectly well. Moreover, I agree that alternatives are far more fundamental structures of our thought than triangles. I nevertheless think that your program will prove inordinately difficult to carry out, for it calls for thought of such abstraction as has never been used before, at least not in physics. I myself would certainly find it too hard, but perhaps the younger generation will take it all in their stride. In short, you and your collaborators must certainly go ahead."

At this point, Elisabeth, who had been listening to us from the other end of the terrace, said: "Do you really believe you can interest the young generation in such difficult problems, problems, moreover, that impinge on the wider context of science and life? From what you have occasionally said about physical research in the great research centers here or in America, it looks very much as if the young generation is almost exclusively preoccupied with details, as if it had placed some sort of taboo on even speaking about a wider context. It's all a bit like astronomy in late antiquity, when the experts were quite content to predict the next eclipse of the sun or the moon from superposed cycles and epicycles, and completely forgot Aristarchus' helocentric system. Might it not happen that interest in your general questions will fade away as well?"

I myself did not feel nearly so pessimistic. "Interest in details is important and necessary," I said, "for, after all, we want to know how things really work. You may remember Niels' oft repeated quotation: 'Clarity is gained through breadth.' As for the taboo, it is not really bad, for it is never intended to prohibit those things of which one must not speak, but to guard them against chatterers and scoffers. Its age-old justification is Goethe's 'Speak, but only to the wise, for the crowd must needs despise.' So the taboo need not really upset us. There will always be young people enough to think about the wider context, if only because they want to be absolutely honest in all things. And that being the case, their number is unimportant."

Those who have thought about Plato's philosophy will know that the world is shaped by images. I would therefore like to conclude this account of conversations over the years with a picture that is indelibly inscribed in my memory. There were four of us—Elisabeth, our two oldest sons and I, and we were driving past blossoming meadows up into hills between Lake Starnberg and Lake Ammer on the way to Seewiesen, where we intended to visit Erich von Holst in the Max Planck Institute for Behavioral Research. Von Holst was not only an excellent biologist, but also a fine viola player and violin maker, and we wanted to ask his advice about a musical instrument. My sons, then young students, had brought along their violin and cello in case there was a chance of playing. Von Holst showed us round his new house, built and furnished most tastefully and imaginatively largely by himself, and led us into a spacious living room into which the sun streamed unobstructed through wide-open windows and French doors. Outside we could see green beeches under a blue sky, and a colorful display of wings by the Institute's feathered charges. Von Holst fetched his viola, sat down between the two young men and joined them in playing the D Major Serenade, a work of Beethoven's youth. It brims over with vital force and joy; faith in the central order keeps casting out faintheartedness and weariness. And as I listened, I grew firm in the conviction that, measured on the human time scale, life, music and science would always go on, even though we ourselves are no more than transient visitors or, in Niels' words, both spectators and actors in the great drama of life.

World Perspectives

What This Series Means

It is the thesis of *World Perspectives* that man is in the process of developing a new consciousness which, in spite of his apparent spiritual and moral captivity, can eventually lift the human race above and beyond the fear, ignorance, and isolation which beset it today. It is to this nascent consciousness, to this concept of man born out of a universe perceived through a fresh vision of reality, that *World Perspectives* is dedicated.

My Introduction to this Series is not of course to be construed as a prefatory essay for each individual book. These few pages simply attempt to set forth the general aim and purpose of the Series as a whole. They try to point to the principle of permanence within change and to define the essential nature of man, as presented by those scholars who have been invited to participate in this intellectual and spiritual movement.

Man has entered a new era of evolutionary history, one in which rapid change is a dominant consequence. He is contending with a fundamental change, since he has intervened in the evolutionary process. He must now better appreciate this fact and then develop the wisdom to direct the process toward his fulfillment rather than toward his destruction. As he learns to apply his understanding of the physical world for practical purposes, he is, in reality, extending his innate capacity and augmenting his ability and his need to communicate as well as his ability to think and to create. And as a result, he is substituting a goal-directed evolutionary process in his struggle against environmental hardship for the slow, but effective, biological evolution which produced modern man through mutation and natural selection. By intelligent intervention in the evolutionary process man has greatly accelerated and greatly expanded the range of his possibilities. But he has not changed the basic fact that it remains a trial and error process, with the danger of taking paths that lead to sterility of mind and heart, moral apathy and intellectual inertia; and even producing social dinosaurs unfit to live in an evolving world.

Only those spiritual and intellectual leaders of our epoch who have a paternity in this extension of man's horizons are invited to participate in this Series: those who are aware of the truth that beyond the divisiveness among men there exists a primordial unitive power since we are all bound together by a common humanity more fundamental than any unity of dogma; those who recognize that the centrifugal force which has scattered and atomized mankind must be replaced by an integrating structure and process capable of bestowing meaning and purpose on existence; those who realize that science itself, when not inhibited by the limitations of its own methodology, when chastened and humbled, commits man to an indeterminate range of yet undreamed consequences that may flow from it.

Virtually all of our disciplines have relied on conceptions which are now incompatible with the Cartesian axiom, and with the static world view we once derived from it. For underlying the new ideas, including those of modern physics, is a unifying order, but it is not causality; it is purpose, and not the purpose of the universe and of man, but the purpose *in* the universe and *in* man. In other words, we seem to inhabit a world of dynamic process and structure. Therefore we need a calculus of potentiality rather than one of probability, a dialectic of polarity, one in which unity and diversity are redefined as simultaneous and necessary poles of the same essence.

Our situation is new. No civilization has previously had to face the challenge of scientific specialization, and our response must be new. Thus this Series is committed to ensure that the spiritual and moral needs of man as a human being and the scientific and intellectual resources at his command for *life* may be brought into a productive, meaningful and creative harmony.

In a certain sense we may say that man now has regained his former geocentric position in the universe. For a picture of the Earth has been made available from distant space, from the lunar desert, and the sheer isolation of the Earth has become plain. This is as new and as powerful an idea in history as any that has ever been born in man's consciousness. We are all becoming seriously concerned with our natural environment. And this concern is not only the result of the warnings given by biologists, ecologists and conservationists. Rather it is the result of a deepening awareness that something new has happened, that the planet Earth is a unique and precious place. Indeed, it may not be a mere coincidence that this awareness should have been born at the exact moment when man took his first step into outer space.

This Series endeavors to point to a reality of which scientific theory has revealed only one aspect. It is the commitment to this reality that lends universal intent to a scientist's most original and solitary thought. By acknowledging this frankly we shall restore science to the great family of human aspirations by which men hope to fulfill themselves in the world community as thinking and sentient beings. For our problem is to discover a principle of differentiation and yet relationship lucid enough to justify and to purify scientific, philosophic and all other knowledge, both discursive and intuitive, by accepting their interdependence. This is the crisis in consciousness made articulate through the crisis in science. This is the new awakening.

Each volume presents the thought and belief of its author and points to the way in which religion, philosophy, art, science, economics, politics and history may constitute that form of human activity which takes the fullest and most precise account of variousness, possibility, complexity and difficulty. Thus *World Perspectives* endeavors to define that ecumenical power of the mind and heart which enables man through his mysterious greatness to re-create his life.

This Series is committed to a re-examination of all those sides of human endeavor which the specialist was taught to believe he could safely leave aside. It attempts to show the structural kinship between subject and object; the indwelling of the one in the other. It interprets present and past events impinging on human life in our growing World Age and envisages what man may yet attain when summoned by an unbending inner necessity to the quest of what is most exalted in him. Its purpose is to offer new vistas in terms of world and human development while refusing to betray the intimate correlation between universality and individuality, dynamics and form, freedom and destiny. Each author deals with the increasing realization that spirit and nature are not separate and apart; that intuition and reason must regain their importance as the means of perceiving and fusing inner being with outer reality.

World Perspectives endeavors to show that the conception of wholeness, unity, organism is a higher and more concrete conception than that of matter and energy. Thus an enlarged meaning of life, of biology, not as it is revealed in the test tube of the laboratory but as it is experienced within the organism of life itself, is attempted in this Series. For the principle of life consists in the tension which connects spirit with the realm of matter, symbiotically joined. The element of life is dominant in the very texture of nature, thus rendering life, biology, a transempirical science. The laws of life have their origin beyond their mere physical manifestations and compel us to consider their spiritual source. In fact, the widening of the conceptual framework has not only served to restore order within the respective branches of knowledge, but has also disclosed analogies in man's position regarding the analysis and synthesis of experience in apparently separated domains of knowledge, suggesting the possibility of an ever more embracing objective description of the meaning of life.

Knowledge, it is shown in these books, no longer consists in a manipulation of man and nature as opposite forces, nor in the reduction of data to mere statistical order, but is a means of liberating mankind from the destructive power of fear, pointing the way toward the goal of the rehabilitation of the human will and the rebirth of faith and confidence in the human person. The works published also endeavor to reveal that the cry for patterns, systems and authorities is growing less insistent as the desire grows stronger in both East and West for the recovery of a dignity, integrity and self-realization which are the inalienable rights of man who may now guide change by means of conscious purpose in the light of rational experience.

The volumes in this Series endeavor to demonstrate that only in a society in which awareness of the problems of science exists, can its discoveries start great waves of change in human culture, and in such a manner that these discoveries may deepen and not erode the sense of universal human community. The differences in the disciplines, their epistemological exclusiveness, the variety of historical experiences, the differences of traditions, of cultures, of languages, of the arts, should be protected and preserved. But the interrelationship and unity of the whole should at the same time be accepted.

The authors of *World Perspectives* are of course aware that the ultimate answers to the hopes and fears which pervade modern society rest on the moral fibre of man, and on the wisdom and responsibility of those who promote the course of its development. But moral decisions cannot dispense with an insight into the interplay of the objective elements which offer and limit the choices made. Therefore an understanding of what the issues are, though not a sufficient condition, is a necessary prerequisite for directing action toward constructive solutions.

Other vital questions explored relate to problems of international understanding as well as to problems dealing with prejudice and the resultant tensions and antagonisms. The growing perception and responsibility of our World Age point to the new reality that the individual person and the collective person supplement and integrate each other; that the thrall of totalitarianism of both left and right has been shaken in the universal desire to recapture the authority of truth and human totality. Mankind can finally place its trust not in a proletarian authoritarianism, not in a secularized humanism, both of which have betrayed the spiritual property right of history, but in a sacramental brotherhood and in the unity of knowledge. This new consciousness has created a widening of human horizons beyond every parochialism, and a revolution in human thought comparable to the basic assumption, among the ancient Greeks, of the sovereignty of reason; corresponding to the great effulgence of the moral conscience articulated by the Hebrew prophets; analogous to the fundamental assertions of Christianity; or to the beginning of the new scientific era, the era of the science of dynamics, the experimental foundations of which were laid by Galileo in the Renaissance.

An important effort of this Series is to re-examine the contradictory meanings and applications which are given today to such terms as democracy, freedom, justice, love, peace, brotherhood and God. The purpose of such inquiries is to clear the way for the foundation of a genuine *world* history not in terms of nation or race or culture but in terms of man in relation to God, to himself, his fellow man and the universe, that reach beyond immediate self-interest. For the meaning of the World Age consists in respecting man's hopes and dreams which lead to a deeper understanding of the basic values of all peoples.

World Perspectives is planned to gain insight into the meaning of man, who not only is determined by history but who also determines history. History is to be understood as concerned not only with the life of man on this planet but as including also such cosmic influences as interpenetrate our human world. This generation is discovering that history does not conform to the social optimism of modern civilization and that the organization of human communities and the establishment of freedom and peace are not only intellectual achievements but spiritual and moral achievements as well, demanding a cherishing of the wholeness of human personality, the "unmediated wholeness of feeling and thought," and constituting a never-ending challenge to man, emerging from the abyss of meaninglessness and suffering, to be renewed and replenished in the totality of his life.

Justice itself, which has been "in a state of pilgrimage and crucifixion" and now is being slowly liberated from the grip of social and political demonologies in the East as well as in the West, begins to question its own premises. The modern revolutionary movements which have challenged the sacred institutions of society by protecting social injustice in the name of social justice are here examined and re-evaluated.

In the light of this, we have no choice but to admit that the *un*freedom against which freedom is measured must be retained with it, namely, that the aspect of truth out of which the night view appears to emerge, the darkness of our time, is as little abandonable as is man's subjective advance. Thus the two sources of man's consciousness are inseparable, not as dead but as living and complementary, an aspect of that "principle of complementarity" through which Niels Bohr has sought to unite the quantum and the wave, both of which constitute the very fabric of life's radiant energy.

There is in mankind today a counterforce to the sterility and danger of a quantitative, anonymous mass culture; a new, if sometimes imperceptible, spiritual sense of convergence toward human and world unity on the basis of the sacredness of each human person and respect for the plurality of cultures. There is a growing awareness that equality may not be evaluated in mere numerical terms but is proportionate and analogical in its reality. For when equality is equated with interchangeability, individuality is negated and the human person transmuted into a faceless mask.

We stand at the brink of an age of a world in which human life presses forward to actualize new forms. The false separation of man and nature, of time and space, of freedom and security, is acknowledged, and we are faced with a new vision of man in his organic unity and of history offering a richness and diversity of quality and majesty of scope hitherto unprecedented. In relating the accumulated wisdom of man's spirit to the new reality of the World Age, in articulating its thought and belief, *World Perspectives* seeks to encourage a renaissance of hope in society and of pride in man's decision as to what his destiny will be.

World Perspectives is committed to the recognition that all great changes are preceded by a vigorous intellectual re-evaluation and reorganization. Our authors are aware that the sin of *hubris* may be avoided by showing that the creative process itself is not a free activity if by free we mean arbitrary, or unrelated to cosmic law. For the creative process in the human mind, the developmental process in organic nature and the basic laws of the inorganic realm may be but varied expressions of a universal formative process. Thus *World Perspectives* hopes to show that although the present apocalyptic period is one of exceptional tensions, there is also at work an exceptional movement toward a compensating unity which refuses to violate the ultimate moral power at work in the universe, that very power upon which all human effort must at last depend. In this way we may come to understand that there exists an inherent independence of spiritual and mental growth which, though conditioned by circumstances, is never determined by circumstances. In this way the great plethora of human knowledge may be correlated with an insight into the nature of human nature by being attuned to the wide and deep range of human thought and human experience.

Incoherence is the result of the present disintegrative processes in education. Thus the need for *World Perspectives* expresses itself in the recognition that natural and man-made ecological systems require as much study as isolated particles and elementary reactions. For there is a basic correlation of elements in nature as in man which cannot be separated, which compose each other and alter each other mutually. Thus we hope to widen appropriately our conceptual framework of reference. For our epistemological problem consists in our finding the proper balance between our lack of an all-embracing principle relevant to our way of evaluating life and in our power to express ourselves in a logically consistent manner.

Our Judeo-Christian and Greco-Roman heritage, our Hellenic tradition, has compelled us to think in exclusive categories. But our *experience* challenges us to recognize a totality richer and far more complex than the average observer could have suspected—a totality which compels him to think in ways which the logic of dichotomies denies. We are summoned to revise fundamentally our ordinary ways of conceiving experience, and thus, by expanding our vision and by accepting those forms of thought which also include nonexclusive categories, the mind is then able to grasp what it was incapable of grasping or accepting before.

In spite of the infinite obligation of men and in spite of their finite power, in spite of the intransigence of nationalisms, and in spite of the homelessness of moral passions rendered ineffectual by the technological outlook, beneath the apparent turmoil and upheaval of the present, and out of the transformations of this dynamic period with the unfolding of a world-consciousness, the purpose of *World Perspectives* is to help quicken the "unshaken heart of well-rounded truth" and interpret the significant elements of the World Age now taking shape out of the core of that undimmed continuity of the creative process which restores man to mankind while deepening and enhancing his communion with the universe.

RUTH NANDA ANSHEN

WORLD PERSPECTIVES

Volumes already published

I	APPROACHES TO GOD	*Jacques Maritain*
II	ACCENT ON FORM	*Lancelot Law Whyte*
III	SCOPE OF TOTAL ARCHITECTURE	*Walter Gropius*
IV	RECOVERY OF FAITH	*Sarvepalli Radhakrishnan*
V	WORLD INDIVISIBLE	*Konrad Adenauer*
VI	SOCIETY AND KNOWLEDGE	*V. Gordon Childe*
VII	THE TRANSFORMATIONS OF MAN	*Lewis Mumford*
VIII	MAN AND MATERIALISM	*Fred Hoyle*
IX	THE ART OF LOVING	*Erich Fromm*
X	DYNAMICS OF FAITH	*Paul Tillich*
XI	MATTER, MIND AND MAN	*Edmund W. Sinnott*
XII	MYSTICISM: CHRISTIAN AND BUDDHIST	*Daisetz Teitaro Suzuki*
XIII	MAN'S WESTERN QUEST	*Denis de Rougemont*
XIV	AMERICAN HUMANISM	*Howard Mumford Jones*
XV	THE MEETING OF LOVE AND KNOWLEDGE	*Martin C. D'Arcy, S.J.*
XVI	RICH LANDS AND POOR	*Gunnar Myrdal*
XVII	HINDUISM: ITS MEANING FOR THE LIBERATION OF THE SPIRIT	*Swami Nikhilananda*
XVIII	CAN PEOPLE LEARN TO LEARN?	*Brock Chisholm*
XIX	PHYSICS AND PHILOSOPHY	*Werner Heisenberg*
XX	ART AND REALITY	*Joyce Cary*
XXI	SIGMUND FREUD'S MISSION	*Erich Fromm*

XXII	MIRAGE OF HEALTH	René Dubos
XXIII	ISSUES OF FREEDOM	Herbert J. Muller
XXIV	HUMANISM	Moses Hadas
XXV	LIFE: ITS DIMENSIONS AND ITS BOUNDS	
		Robert M. MacIver
XXVI	CHALLENGE OF PSYCHICAL RESEARCH	Gardner Murphy
XXVII	ALFRED NORTH WHITEHEAD: HIS REFLECTIONS ON MAN AND NATURE	Ruth Nanda Anshen
XXVIII	THE AGE OF NATIONALISM	Hans Kohn
XXIX	VOICES OF MAN	Mario Pei
XXX	NEW PATHS IN BIOLOGY	Adolf Portmann
XXXI	MYTH AND REALITY	Mircea Eliade
XXXII	HISTORY AS ART AND AS SCIENCE	H. Stuart Hughes
XXXIII	REALISM IN OUR TIME	Georg Lukács
XXXIV	THE MEANING OF THE TWENTIETH CENTURY	
		Kenneth E. Boulding
XXXV	ON ECONOMIC KNOWLEDGE	Adolph Lowe
XXXVI	CALIBAN REBORN	Wilfred Mellers
XXXVII	THROUGH THE VANISHING POINT	
		Marshall McLuhan and Harley Parker
XXXVIII	THE REVOLUTION OF HOPE	Erich Fromm
XXXIX	EMERGENCY EXIT	Ignazio Silone
XL	MARXISM AND THE EXISTENTIALISTS	Raymond Aron
XLI	PHYSICAL CONTROL OF THE MIND	
		José M. R. Delgado, M.D.
XLII	PHYSICS AND BEYOND	Werner Heisenberg
XLIII	ON CARING	Milton Mayeroff
XLIV	DESCHOOLING SOCIETY	Ivan Illich
XLV	REVOLUTION THROUGH PEACE	Dom Hélder Câmara

About the Author

Werner Heisenberg was born in Würzburg, Germany, in 1901. He was educated at the Universities of Munich and Göttingen and in 1932 was awarded the Nobel Prize for his work in theoretical atomic physics. He is now Director of the Max Planck Institute for Physics and Astrophysics in Munich.

About the Editor of This Series

Ruth Nanda Anshen, philosopher and editor, plans and edits *World Perspectives, Religious Perspectives, Credo Perspectives, Perspectives in Humanism* and *The Science of Culture Series.* She also writes and lectures on the relationship of knowledge to the nature and meaning of man and existence.

harper 🔥 torchbooks

American Studies: General

CARL N. DEGLER: Out of Our Past: *The Forces that Shaped Modern America* CN/2
ROBERT L. HEILBRONER: The Limits of American Capitalism TB/1305
JOHN HIGHAM, Ed.: The Reconstruction of American History TB/1068
JOHN F. KENNEDY: A Nation of Immigrants. *Illus. Revised and Enlarged. Introduction by Robert F. Kennedy* TB/1118
GUNNAR MYRDAL: An American Dilemma: *The Negro Problem and Modern Democracy. Introduction by the Author.*
Vol. I TB/1443; Vol. II TB/1444
GILBERT OSOFSKY, Ed.: The Burden of Race: *A Documentary History of Negro-White Relations in America* TB/1405
ARNOLD ROSE: The Negro in America: *The Condensed Version of Gunnar Myrdal's An American Dilemma* TB/3048

American Studies: Colonial

BERNARD BAILYN: The New England Merchants in the Seventeenth Century TB/1149
ROBERT E. BROWN: Middle-Class Democracy and Revolution in Massachusetts, 1691–1780. *New Introduction by Author* TB/1413
JOSEPH CHARLES: The Origins of the American Party System TB/1049

American Studies: The Revolution to 1900

GEORGE M. FREDRICKSON: The Inner Civil War: *Northern Intellectuals and the Crisis of the Union* TB/1358
WILLIAM W. FREEHLING: Prelude to Civil War: *The Nullification Controversy in South Carolina, 1816–1836* TB/1359
HELEN HUNT JACKSON: A Century of Dishonor: *The Early Crusade for Indian Reform.* ‡ *Edited by Andrew F. Rolle* TB/3063
RICHARD B. MORRIS, Ed.: Alexander Hamilton and the Founding of the Nation. *New Introduction by the Editor* TB/1448
RICHARD B. MORRIS: The American Revolution Reconsidered TB/1363
GILBERT OSOFSKY, Ed.: Puttin' On Ole Massa: *The Slave Narratives of Henry Bibb, William Wells Brown, and Solomon Northup* ‡ TB/1432

American Studies: The Twentieth Century

WILLIAM E. LEUCHTENBURG: Franklin D. Roosevelt and the New Deal: 1932-1940. † *Illus.* TB/3025
WILLIAM E. LEUCHTENBURG, Ed.: The New Deal: *A Documentary History* + HR/1354

Asian Studies

WOLFGANG FRANKE: China and the West: *The Cultural Encounter, 13th to 20th Centuries. Trans. by R. A. Wilson* TB/1326
L. CARRINGTON GOODRICH: A Short History of the Chinese People. *Illus.* TB/3015
BENJAMIN I. SCHWARTZ: Chinese Communism and the Rise of Mao TB/1308

Economics & Economic History

PETER F. DRUCKER: The New Society: *The Anatomy of Industrial Order* TB/1082
ROBERT L. HEILBRONER: The Great Ascent: *The Struggle for Economic Development in Our Time* TB/3030
W. ARTHUR LEWIS: The Principles of Economic Planning. *New Introduction by the Author*° TB/1436

Historiography and History of Ideas

J. BRONOWSKI & BRUCE MAZLISH: The Western Intellectual Tradition: *From Leonardo to Hegel* TB/3001
WILHELM DILTHEY: Pattern and Meaning in History: *Thoughts on History and Society.*° *Edited with an Intro. by H. P. Rickman* TB/1075
J. H. HEXTER: More's Utopia: *The Biography of an Idea. Epilogue by the Author* TB/1195
ARTHUR O. LOVEJOY: The Great Chain of Being: *A Study of the History of an Idea* TB/1009

History: Medieval

F. L. GANSHOF: Feudalism TB/1058
DENYS HAY: The Medieval Centuries ° TB/1192
HENRY CHARLES LEA: A History of the Inquisition of the Middle Ages. || *Introduction by Walter Ullmann* TB/1456

† The New American Nation Series, edited by Henry Steele Commager and Richard B. Morris.
‡ American Perspectives series, edited by Bernard Wishy and William E. Leuchtenburg.
α History of Europe series, edited by J. H. Plumb.
§ The Library of Religion and Culture, edited by Benjamin Nelson.
‖ Researches in the Social, Cultural, and Behavioral Sciences, edited by Benjamin Nelson.
Σ Harper Modern Science Series, edited by James R. Newman.
° Not for sale in Canada.
+ Documentary History of the United States series, edited by Richard B. Morris.
# Documentary History of Western Civilization series, edited by Eugene C. Black and Leonard W. Levy.
A The Economic History of the United States series, edited by Henry David et al.
¶ European Perspectives series, edited by Eugene C. Black.
** Contemporary Essays series, edited by Leonard W Levy.
* The Stratum Series, edited by John Hale.

History: Renaissance & Reformation

JACOB BURCKHARDT: The Civilization of the Renaissance in Italy. *Introduction by Benjamin Nelson and Charles Trinkaus. Illus.*
Vol. I TB/40; Vol. II TB/41
JOEL HURSTFIELD: The Elizabethan Nation
TB/1312
ALFRED VON MARTIN: Sociology of the Renaissance. ° *Introduction by W. K. Ferguson*
TB/1099
J. H. PARRY: The Establishment of the European Hegemony: 1415-1715: *Trade and Exploration in the Age of the Renaissance* TB/1045

History: Modern European

MAX BELOFF: The Age of Absolutism, 1660-1815
TB/1062
ALAN BULLOCK: Hitler, A Study in Tyranny. ° *Revised Edition. Illus.* TB/1123
JOHANN GOTTLIEB FICHTE: Addresses to the German Nation. *Ed. with Intro. by George A. Kelly* ¶ TB/1366
H. STUART HUGHES: The Obstructed Path: *French Social Thought in the Years of Desperation* TB/1451
JOHAN HUIZINGA: Dutch Cviilization in the 17th Century and Other Essays TB/1453
JOHN MCMANNERS: European History, 1789-1914: *Men, Machines and Freedom* TB/1419
FRANZ NEUMANN: Behemoth: *The Structure and Practice of National Socialism, 1933-1944*
TB/1289
A. J. P. TAYLOR: From Napoleon to Lenin: *Historical Essays* ° TB/1268
H. R. TREVOR-ROPER: Historical Essays TB/1269

Philosophy

HENRI BERGSON: Time and Free Will: *An Essay on the Immediate Data of Consciousness* °
TB/1021
G. W. F. HEGEL: Phenomenology of Mind. ° ‖ *Introduction by George Lichtheim* TB/1303
H. J. PATON: The Categorical Imperative: *A Study in Kant's Moral Philosophy* TB/1325
MICHAEL POLANYI: Personal Knowledge: *Towards a Post-Critical Philosophy* TB/1158
LUDWIG WITTGENSTEIN: The Blue and Brown Books ° TB/1211
LUDWIG WITTGENSTEIN: Notebooks, 1914-1916
TB/1441

Political Science & Government

C. E. BLACK: The Dynamics of Modernization: *A Study in Comparative History* TB/1321
DENIS W. BROGAN: Politics in America. *New Introduction by the Author* TB/1469
KARL R. POPPER: The Open Society and Its Enemies *Vol. I: The Spell of Plato* TB/1101
Vol: II: The High Tide of Prophecy: Hegel, Marx, and the Aftermath TB/1102
CHARLES SCHOTTLAND, Ed.: The Welfare State **
TB/1323
JOSEPH A. SCHUMPETER: Capitalism, Socialism and Democracy TB/3008
PETER WOLL, Ed.: Public Administration and Policy: *Selected Essays* TB/1284

Psychology

LUDWIG BINSWANGER: Being-in-the-World: *Selected Papers.* ‖ *Trans. with Intro. by Jacob Needleman* TB/1365

MIRCEA ELIADE: Cosmos and History: *The Myth of the Eternal Return* § TB/2050
SIGMUND FREUD: On Creativity and the Unconscious: *Papers on the Psychology of Art, Literature, Love, Religion.* § *Intro. by Benjamin Nelson* TB/45
J. GLENN GRAY: The Warriors: *Reflections on Men in Battle. Introduction by Hannah Arendt* TB/1294
WILLIAM JAMES: Psychology: *The Briefer Course. Edited with an Intro. by Gordon Allport* TB/1034

Religion

TOR ANDRAE: Mohammed: *The Man and his Faith* TB/62
KARL BARTH: Church Dogmatics: *A Selection. Intro. by H. Hollwitzer. Ed. by G. W. Bromiley* TB/95
NICOLAS BERDYAEV: The Destiny of Man TB/61
MARTIN BUBER: The Prophetic Faith TB/73
MARTIN BUBER: Two Types of Faith: *Interpenetration of Judaism and Christianity*
TB/75
RUDOLF BULTMANN: History and Eschatalogy: *The Presence of Eternity* TB/91
EDWARD CONZE: Buddhism: *Its Essence and Development. Foreword by Arthur Waley*
TB/58
H. G. CREEL: Confucius and the Chinese Way
TB/63
FRANKLIN EDGERTON, Trans. & Ed.: The Bhagavad Gita TB/115
M. S. ENSLIN: Christian Beginnings TB/5
M. S. ENSLIN: The Literature of the Christian Movement TB/6
HENRI FRANKFORT: Ancient Egyptian Religion: *An Interpretation* TB/77
IMMANUEL KANT: Religion Within the Limits of Reason Alone. *Introduction by Theodore M. Greene and John Silber* TB/67
GABRIEL MARCEL: Homo Viator: *Introduction to a Metaphysics of Hope* TB/397
H. RICHARD NIEBUHR: Christ and Culture TB/3
H. RICHARD NIEBUHR: The Kingdom of God in America TB/49
SWAMI NIKHILANANDA, Trans. & Ed.: The Upanishads TB/114
F. SCHLEIERMACHER: The Christian Faith. *Introduction by Richard R. Niebuhr.*
Vol. I TB/108 Vol. II TB/109

Sociology and Anthropology

KENNETH B. CLARK: Dark Ghetto: *Dilemmas of Social Power. Foreword by Gunnar Myrdal*
TB/1317
KENNETH CLARK & JEANNETTE HOPKINS: A Relevant War Against Poverty: *A Study of Community Action Programs and Observable Social Change* TB/1480
GARY T. MARX: Protest and Prejudice: *A Study of Belief in the Black Community* TB/1435
ROBERT K. MERTON, LEONARD BROOM, LEONARD S. COTTRELL, JR., Editors: Sociology Today: *Problems and Prospects* ‖
Vol. I TB/1173; Vol. II TB/1174
GILBERT OSOFSKY: Harlem: The Making of a Ghetto: *Negro New York, 1890-1930* TB/1381
PHILIP RIEFF: The Triumph of the Therapeutic: *Uses of Faith After Freud* TB/1360
GEORGE ROSEN: Madness in Society: *Chapters in the Historical Sociology of Mental Illness.* ‖ *Preface by Benjamin Nelson* TB/1337